AF880144

Number: The Categorical View

A Companion Volume to *Number and the Nature of Mathematics*

Joseph Bergin, Ph.D.

Published by Slant Flying Press, 2023

ISBN: 978-1-940113-08-1

Acknowledgements:

Thanks to Ian Benson of esociality.com for getting me to think about this project and providing guidance and feedback along the way. He also provided the insight that led to the discussion of the Distributive Law, though his expression was different.

Cover photography by the author.

Some figures taken from https://openclipart.org.

Some figures taken from WikiCommons.

London Tube Map taken from Transport of London.

Brooklyn Bridge Photo by Jerome Dominici:
https://www.pexels.com/photo/gray-bridge-297303/

The name of the song is called 'Haddocks' Eyes.'"

"Oh, that's the name of the song, is it?". Alice said, trying to feel interested.

"No, you don't understand," the knight said, looking a little vexed. That's what the name is called. The name really is 'The Aged, Aged Man.'"

"Then I ought to have said 'That's what the song is called'?" Alice corrected herself.

"No, you oughtn't: thats quite another thing! The song is called 'Ways and Means': but that's only what it is called, you know!"

"Well, what is the song, then?" said Alice, who was, by this time completely bewildered.

"I was coming to that," the Knight said. "The song really is 'A-sitting On a Gate': and the tune's my own invention."

- Lewis Carrol, Through the Looking Glass

Introduction

When we refer to *The Book* we mean *Number and the Nature of Mathematics* by the same author. As in The Book we will mark certain terms with (+) to indicate that the term might profitably be looked up in Wikipedia.

The Book gave a traditional and historical view of the development of Number and some of the relationships between different traditional number systems. In this book we take a somewhat different, more modern view; that of Category *Theory,* which is surprisingly simple, but which has wide application to mathematics, and not just number. Many modern Algebraic systems, including Algebraic Topology(+) and Algebraic Geometry(+) have a structure layered on top of categories. But those are far beyond what we will see here.

A note on terminology here. In The Book we started with a definition of the Natural Numbers that began with 1, not 0. That was for two reasons. First is that it was the original Peano formulation and second it demonstrated the awkwardness of the ad-hoc addition of 0 to form the Whole Numbers. Modern usage usually has the Natural Numbers starting with 0, but we will keep the original (from The Book) terminology here as it will allow us to make a few distinctions as well as to keep to the idea that this is only a companion to the earlier book.

While we discuss Category Theory here, this is not a complete treatment. In particular, we only look at *small* categories in any detail; those that can be described within a rigorous set theory.

The book closes with a brief introduction of how category theory applies to computing and how that circles back to Number.

Audience:

As with The Book, anyone who has studied an introduction to Calculus should be able, with a bit of work, to follow what we do here. But you will probably need a copy of *Number and the Nature of Mathematics*: ISBN 978-1-940113-07-4.

Table of Contents

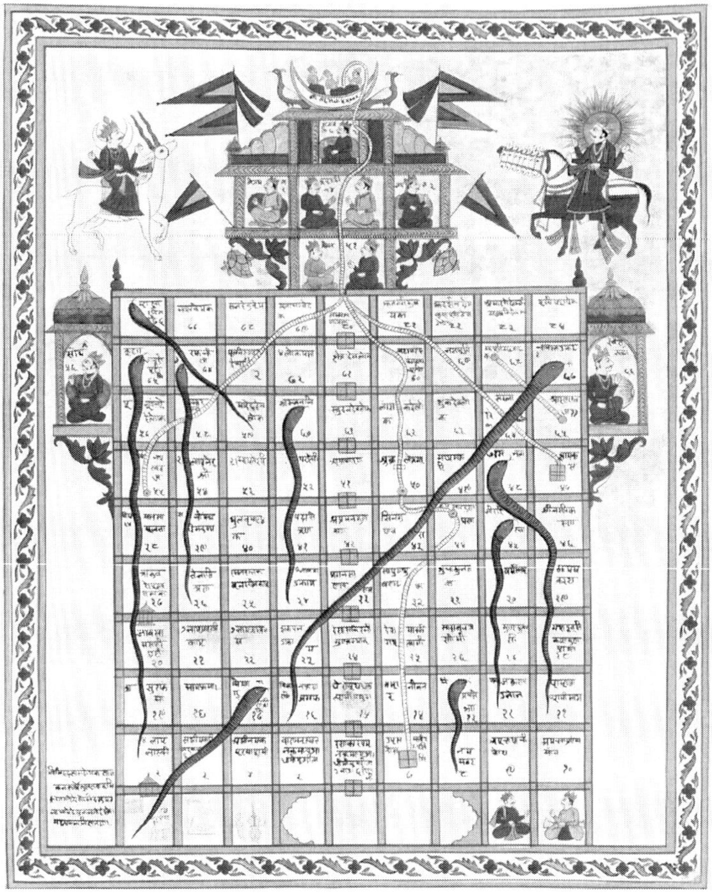

1 Fundamentals

Category Theory

A *category* is a generalized mathematical structure that abstracts several common elements found throughout mathematics. As such, it is sort of the mathematical theory of mathematics, or a meta-level approach to mathematics.

A Category, $\mathcal{C}$, consists of

A collection of "objects", Ob($\mathcal{C}$).

A collection of one directional "arrows" between objects: Hom($\mathcal{C}$).

Such that:

For each object, *a,* in Ob($\mathcal{C}$), there is at least one *arrow* from a to a in Hom($\mathcal{C}$), called the *identity* arrow: I_a.

Arrows may be composed: If there is an arrow, f, from a to b and an arrow, g, from b to c, then there is an arrow from a to c, called the *composition*, written g∘f, or, more simply, just gf.

With:

The identity arrows behave as follows: If f is an arrow from a to b then $f \circ I_a = f = I_b \circ f$.

(more...)

Composition of arrows is *associative*: If f is an arrow from a to b, g is an arrow from b to c, and h is an arrow from c to d, then $(h \circ g) \circ f = h \circ (g \circ f)$.

Note that the "collections" need not be sets. They might not satisfy classic set theory. We will restrict the following to the situation where they are sets, however. There is a more general *class theory*(+) of collections that extends set theory and avoids some of the paradoxes that arise in naïve set theory. There is also an extension of naïve set theory (Zermelo-Fraenkel(+) set theory) that avoids the paradoxes, but Category Theory permits even more general collections.

The arrows are more formally called *morphisms* and the metaphor for morphisms is something like transformations or functions, but possibly much more general. The collections might be empty, though if there are any objects, then there must be identity arrows at least. There might be various interpretations given to morphisms, depending on the particular case. An arrow might represent a *modification*, such as the rotation of a figure, or some *relationship* between objects. If f is an arrow from *a* to *b*, then a is called the *source* (alternately *domain*) of f and b is called the *destination* (co*domain*) of f. In general, there might be several arrows from any given a to any given b, or none at all. The collection of morphisms in $\mathcal{C}$ with source a and destination b is named $\mathcal{C}(a, b)$. $\mathcal{C}(a, a)$ contains at least I_a. And, if f is a morphism from a to b, and g is a morphism from b to c, then the composition $g \circ f$ is in $\mathcal{C}(a, c)$.

Here the *equality* of arrows is taken to mean "has the same meaning as" or "produces the same result as", or "carries the same information or data as". So, it might be that two objects have several arrows connecting them, each with a different meaning. For example, in the Rational Numbers we might think of a +3 arrow connecting objects 3 and 6 but also a *2 arrow also connecting those objects. But we might also find that such arrows carry the same information. We will explore the technical meaning of "same" later: isomorphism.

There isn't a universal identity arrow for the category, rather there is an identity arrow associated with each object. It can be different for each object, or not, as the examples will show.

Function notation is often used to represent arrows. If f is an arrow from a to b we might write f(a) = b or f:a→b. Be careful with the notation, since f might only apply to a single object, a, not the entire category. But function notation explains the order of composition: If *f* is an arrow from a to b and *g* is an arrow from b to c then $g(f(a)) = g(b) = c = g \circ f(a)$.

Graphical Representation of the Rules

The rules for categories can be depicted graphically as follows.

An arrow f from an object a to an object b might be depicted graphically as:

An identity arrow I_a for object *a* is either:

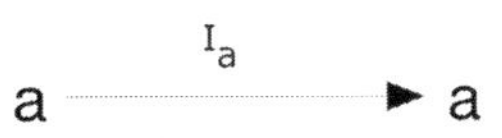

or:

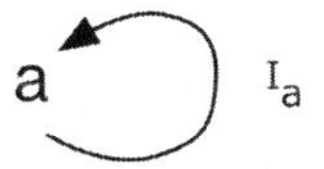

The composition of two arrows, showing the result as well is:

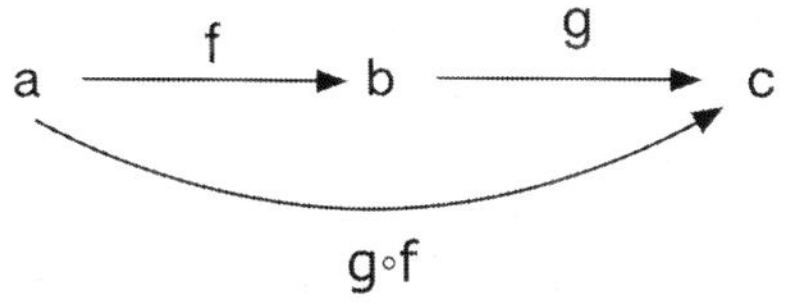

The associative rule for the morphisms of a category looks like the following:

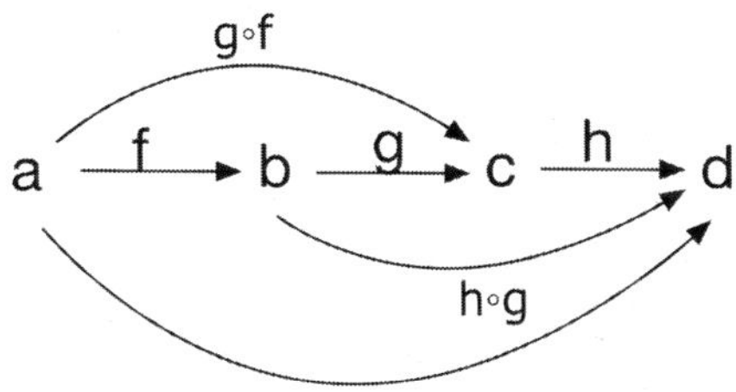

$$h \circ (g \circ f) = (h \circ g) \circ f$$

Observe that the identity arrow for an object is *unique* for that object. If there were two identity morphisms, I and J, then the identity property of J implies that I = IJ, and the identity property of I implies that IJ = J also. So I = J. Also note that, when we have associativity, instead of writing h(gf) or (hg)f we can just write hgf as we normally do with associative relationships, such as addition, in mathematics.

Simple Examples of Categories

The simplest category is simply empty: no objects, no morphisms. It doesn't seem very interesting, but note that the definition of Category is such that we only need arrows if we have objects. This category might be called *EMPTY*, alternately, *ZERO*.

A single object and a single identity arrow is enough to make a category. Most of the requirements are automatically fulfilled. The object doesn't even need any real interpretation. This category might be called *SELF*, alternately, *ONE*. A category with only one object is called a monoid, though a monoid can have several arrows, as we shall see.

A more interesting category has the subsets of some fixed, usually non-empty, set S as objects. There is an arrow from set a to set b provided a is a subset of b. Note that a set is always a subset of itself, providing identity arrows. If S is empty this is just SELF (not EMPTY). And, if S has more than one element then there will be some pairs of objects with arrows between them and others without. As an exercise, verify the other properties for this category. I'll call this category *SUBSETS(S)*.

There is a subtlety, here, however. The question is, how many identity arrows do we have. Technically, the answer is that we have a separate arrow for each object. But they all seem to be the same: a set is a subset of itself. But, since the source of an arrow is part of the concept, there is a distinct, if similar, identity morphism for each object. This same thing occurs in other categories with more than one object. And not every arrow from an object to itself is an identity arrow. An identity morphism also needs to compose properly with other arrows on both the left and right, leaving the effect of the other arrow unchanged.

We verify the associativity rule for this category by looking at the case of four subsets (not necessarily distinct) and three subset relationships. Suppose $A \subseteq B \subseteq C \subseteq D$. Here we use the same symbol, but there are actually three morphisms f(A) = B, g(B) = C, and h(C) = D. We need to verify that $(h \circ g) \circ f = h \circ (g \circ f)$. First, looking at the left side, we see that $(h \circ g) \circ f(A) = h \circ g(B)$ since only f has been applied. But $h \circ g(B) = h(C)$ applying morphism g since $h \circ g(x) = h(g(x))$. And h(C) = D, so the left hand side, $(h \circ g) \circ f$, evaluates to D and we conclude that $A \subseteq D$ using the conventional notation. The reader can similarly show that the right hand side, $h \circ (g \circ f)$, applied to A also results in D, yielding the rule of associativity.

Let A be a Natural Number and consider the set of exact factors of A (not just prime factors). For example, if A = 30 then its factors are {1, 2, 3, 5, 6, 10, 15, 30}. These are the objects. There is an arrow from a to b provided that a is a divisor of b. So there is an arrow from 3 to 15, for example, but not from 3 to 5. Every Natural Number divides itself, so we have identity arrows. Composition is just the statement that if a divides b and b divides c, then a divides c as well. Call this category *FACTORS*(30). What is FACTORS(1)?

As with many categories, you need to be able to interpret the meaning of composed arrows. This will be especially important when verifying associativity. Suppose, in the above case, that we want to use a vertical bar to indicate a morphism. So, 2 divides 6 becomes 2 | 6, which is common mathematical notation. In the category, this means we have an arrow, or morphism from 2 to 6. But (2 | 6) | 30 would not be common notation in math, since 2 | 6 results in "true" and "*true* | 30" makes no sense. But, in the sense of a category, an arrow results in the object *pointed to* so, as a morphism 2 | 6 has a result of 6. This makes sense out of the composition: (2 | 6) | 30, which we can also write as 2 | 6 | 30.

When drawing the set of arrows for a category you can take some shortcuts because of the nature of a category. Below is the graph for FACTORS(12).

Note that there is no need to draw the identity arrows since we assume they are always present for every object. Also there is no need to draw a direct arrow from 2 to 12 since composition guarantees its presence. What is the characterization of prime numbers in such a diagram? Composites?

As an exercise, draw the arrows for FACTORS(30). It is a bit messier. FACTORS(36) is cleaner.

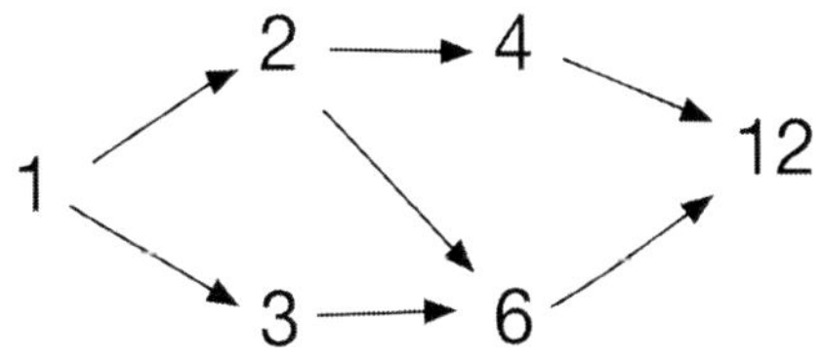

Note that there are three "pathways" to get from 1 to 12 in this diagram. All of them are compositions. The question arises, then, how many morphisms are there from 1 to 12. The answer is, actually, just one, though it takes work to show that all of these pathways result in the *same* morphism. The three pathways carry the same "data", namely 1 is an exact divisor of 12. We will need to explore what it means for morphisms to be the same. We need to look at Arrow Categories to do this, and will, later. Factors(12) actually has 18 distinct arrows. Here, and in what follows, "pathway" is just a synonym for morphism, often a composition.

Suppose we drop the "object" 1 from the above, just considering the resulting graph. It is still a category, but there would be separate morphisms with destination 12, one with source 2 and the other with source 3. But there is still only one from 2 to 12, even though there are two "pathways".

Here is an alternate view of FACTORS(12) with the arrows labelled. The information carried by each is really the same for all; the source exactly divides the destination.

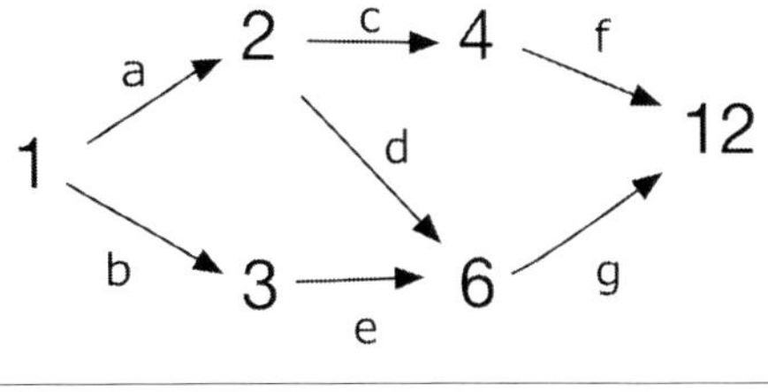

And here is that same diagram in a more "category-like" form.

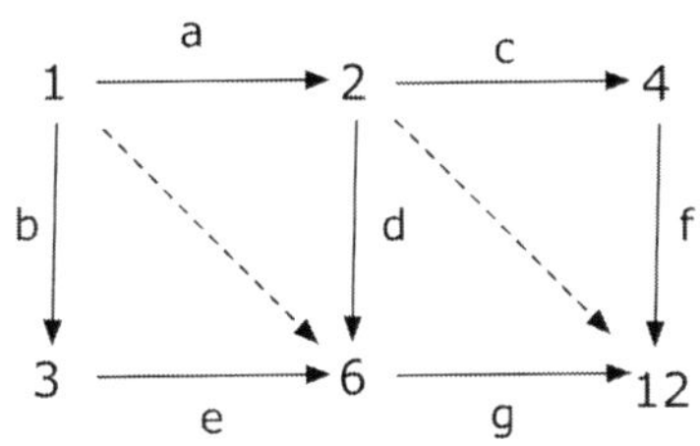

Note that if we follow the a and then the d arrows we wind up at the same place, 6, as if we follow the b and then the e arrow. We say that this part of the diagram is commutative or "forms a *square*". In fact there are other commutative parts as well as the whole. So, fca and geb form part of a commutative diagram as well. These ideas are important in the sequel.

Note, also, that if we replace the names of the objects in FACTORS(12) with something else, say {Rudolph, Edward, Leonard, Ferdinand, Caspar, Catherine}, but keep the arrows the same, then we have the same category structurally, though the interpretation of the arrows would be different - or totally abstract. It is the *structure* that makes the category.

Partially ordered sets form categories. A partially ordered set (*poset*), X, is a set with a "comparison" operation, usually denoted by ≤, with the following properties:

If x is an element of x then x≤x (reflexivity).
If x and y are elements of X and both x≤y and y≤x, then x=y (antisymmetry).
If x, y, and z are elements of X and both x≤y and y≤z, then x≤z (transitivity).

It is the reflexive and transitive rules that imply that a poset is a category. Note also, that subsets of a given set form a poset using the subset relation.

Note that it isn't necessary that the operation be applicable to every pair. That is what makes it "partial". In particular FACTORS(12) is a poset but 2 can't be compared with 3, nor 3 with 4. It is only the morphisms of FACTORS(12) that define the "comparison".

Consider a clock-like structure, $\mathcal{C}$, say with four *hours*: 1, 2, 3, and 4. Consider it as a monoid so the only object is $\mathcal{C}$ itself. There are two kinds of "elementary" arrows. The first is the identity on $\mathcal{C}$, namely I_C. Think of it as advancing zero hours. The second is the process of advancing one hour, the *successor* operation, actually. Note that the successor of 4 on such a clock is 1. Let $h = h^1$ be the arrow that advances $\mathcal{C}$ from wherever it is by one hour. Note that $I_C \circ h = h \circ I_C = h$. If we apply h more than once we can use a shorthand notation: $h \circ h = h^2$, $h \circ h \circ h = h^3$, etc. Note that then $h^4 = I_C$ and $h^5 = h$. We might use h^0 for I_C as well. Moreover we get $h^3 \circ h^2 = h^5 = h$ since the clock "wraps around". Each application of h to $\mathcal{C}$ advances the clock an hour and four such applications wraps it around to the start "time". Call this monoid CLOCK(4). Technically there are an infinite number of arrows (morphisms), but only four distinct ones.

The CLOCK monoid has another interesting property. Think of turning the clock back an hour or more. Then h^{-1} is turning the clock back by one hour. This would be a new kind of morphism, and so it might result in a different category. But now, if we have any ordinary arrow, h^k, then there is a corresponding reverse arrow, say h^{-k} that "undoes" the original under composition, resulting in h^0. Thus every arrow has an "inverse". This arrow structure is $\mathfrak{N}^4$, in fact: the Natural Numbers mod 4. However, the arrows with negative superscripts are redundant (the symbols are redundant) since $h^3 = h^{-1}$ in fact, and similarly for the others. That is to say, -1 = 3 (mod 4). We conclude, then, that this isn't a different category.

I own a mechanical wall clock that permits me to move it forward manually by rotating the hands. But the mechanism won't let me turn it back that way. Thus, in the US in the autumn, when we "fall back" at the end of daylight savings time, I need to turn the clock forward eleven hours, with the same result as turning it back one hour.

Note what this monoid stresses and what it does not. Normally we think of the hour labels on a clock as important, but here they are irrelevant. We could label them bird, cat, dog, elephant if we like and the effect would be the same. This view stresses the transitions (morphisms). Thus we have abstracted away certain details to emphasize others. This is what makes it math, and what raises it to a higher level. Also note that this monoid stresses the passage of time, not the position of the hands on the clock. So h^{17} might represent the passage of 17 hours, but its effect on the clock is the same as h^1.

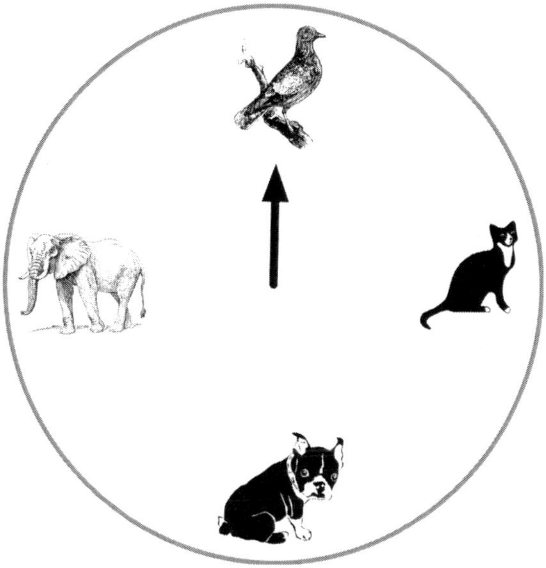

A monoid, having a single object, might best be thought of as something other than a set. It is a "thing" with "states". For a clock, the states can be numbers as on a normal clock or some other indicators. The morphisms represent *transitions* between the states. This is, then, just a change of viewpoint of a concept, which can be fruitful.

A monoid in which every arrow has an inverse is called a *Group*^(+)^. In this example the elements of the group are $\{h^0, h^1, h^2, h^3\}$, the operation is addition modulo 4, or composition of the elements in the category view, and the identity is h^0. You could relabel the elements {+0, +1, +2, +3}, of course, as we shall see below. This group is called the Integers mod 4.

There is a different view, also a category, for a clock-like structure. Let the objects be {1, 2, 3, 4}, or {bird, cat, dog, elephant} if you like, so it isn't a monoid. We will need identity arrows for each of these. So I_4 is the identity arrow for object 4, etc. The identity arrows don't advance the "time" on the clock. We also want transition arrows as in the first view so we define a set of arrows f that act like a function: $f_1 = f(1) = 2$ for the arrow that points from 1 to 2. Likewise we have $f_2 = f(2) = 3$, $f_3 = f(3) = 4$, and $f_4 = f(4) = 1$. Each of these "advances the clock by an hour but specifies the start and end times. These aren't compositions as we used above. We can now compose these arrows so that f_2 followed by f_3 is an arrow from 2 to 4. Note that f_2 can't be composed with f_4 since the destination of the first isn't the source of the second. This view gives some "substance" to the hours themselves, but is fundamentally messier. And messy isn't a real goal in mathematics. But it is a category, in fact. The new identity arrows don't really add anything and the compositions "feel" a bit simpler and cleaner in the original version.

Note that the first view clearly indicates that the effect of an arrow is independent of the start time. Thus, the arrow is applied to the single object, the category, implying that these arrows are independent of any internal structure that the "clock" might have. It isn't obvious that this is the case for the second view.

Another way to look at the above "clock" is in a table of morphisms applied to objects:

	+0	+1	+2	+3
bird	bird	cat	dog	elephant
cat	cat	dog	elephant	bird
dog	dog	elephant	bird	cat
elephant	elephant	bird	cat	dog

In this representation, which merges ideas from the two versions, some names for the arrows are at the top, the source of each arrow is at the left and the destination is a cell in the body of the table. We have also renamed the morphism here, so h^2 from the first version becomes +2, etc.

Note the patterns here, in the vertical patterns of the destinations, the horizontal patterns and the diagonal patterns. Look along diagonals from upper right toward lower left for example. The passage of "time" in this table is movement to the right from any cell, wrapping around if you fall off the right end. And, moving backwards by an "hour" has the same effect as moving forward by three cells. This is the inverse of an arrow. What are the inverses of the other morphisms?

New Categories From Old

One can also create new categories from old. For example, if you create a new thing with objects the same as those of a given category, but with arrow all pointing in the opposite direction of the original arrows, then the new thing is also a category. This "opposite category" or "dual" of a category $\mathcal{C}$, written $\mathcal{C}^{OP}$. The identity arrows of $\mathcal{C}$ and $\mathcal{C}^{OP}$ are the same.

Another interesting idea is that if you have two categories, $\mathcal{C}$ and $\mathcal{D}$, you can form their Cartesian Product $\mathcal{C} \times \mathcal{D}$; ordered pairs with the first element an object of $\mathcal{C}$ and the second an object from $\mathcal{D}$. These are the objects of the new category. The morphisms are likewise ordered-pairs, but of morphisms; if $f:a \rightarrow b$ is a morphism in $\mathcal{C}$ and $g:c \rightarrow d$ one from $\mathcal{D}$, then (a, c), (a, d), (b, c) and (b, d) are in $\mathcal{C} \times \mathcal{D}$ and (f, g) is a morphism taking (a, c) to (b, d). An identity in $\mathcal{C} \times \mathcal{D}$ is the pair made from identities in $\mathcal{C}$ and $\mathcal{D}$. In particular we can form a category from a single one using the same idea; $\mathcal{C} \times \mathcal{C}$, denoted $\mathcal{C}^2$. And we can do the same with more than two, of course. Cartesian Products will appear later.

Monoids: Alternate View

There is another, older, definition of a monoid, from abstract algebra(+), that requires investigation. Suppose we have a set, Y, and a binary operation on Y, say, $f:Y \times Y \rightarrow Y$. The Natural Numbers with multiplication as the operation, for example. We will use + as a symbol for this operation, but it is just a placeholder. Suppose also that the operation is associative: x + (y + z) = (x + y) + z. Finally, suppose there is an identity element, denoted by 0; x + 0 = 0 + x = x, for any x. Then Y, together with the operation is a monoid.

The Whole Numbers with addition are a monoid in this sense. The same is true of the Natural Numbers with multiplication.

Set Y, with such an operation, is traditionally called a monoid. Note that the elements of Y, such as 3 and 42 are values, not *objects* in the category sense. In this view, the operation, +, isn't exactly an arrow in that sense either. Moreover, the identity is unique. How are these two notions related? It will take some thought to get your mind around this, I promise.

Shifting to a category view, don't think of the elements of Y as objects, which would seem natural. Instead, consider them to be *arrows*. So, 5, in the Whole Numbers considered as a category is an arrow with the Whole Numbers as both source and destination.

Next, consider the *operation* on elements of the set to be *composition* of these arrows. So 5 + 3 is the composition of two arrows; the 5 arrow composed with the 3 arrow. In category-speak it is $3 \circ 5$, in fact; first "do" 5, then 3. With this interpretation, The Natural Numbers with multiplication as the operation are a monoid in the category sense: a category with a single object. Note that 1 is the identity arrow here since $3 \circ 1$ and $1 \circ 3$ are just the 3 arrow.

In the traditional view, the identity is unique. There is only one 1. But in the category view there is an identity for every object. But since it is a category with only one object, it isn't that different after all.

This new interpretation of monoids also explains how a monoid with inverses is a group, though the Natural Numbers with multiplication don't qualify. How about the Whole Numbers with addition? Integers? Rational Numbers?

Algebraic Structures, Homomorphisms, Isomorphisms

An *Algebraic Structure* is a non-empty collection, usually a set, of *values* along with a collection of *operations*, usually unary and binary operations, together with a set of *rules* (axioms) that govern the behavior of the operations. The Integers with sum and product as the operations is an example, where the rules are as you have seen them in The Book. Category theory focuses on the structure of such a thing; how the rules enforce behavior.

A mapping from one Algebraic Structure to another might preserve the structure or not. A random mapping, for example, preserves nothing in general. In order to preserve structure, a mapping f:A → B needs to map values of A to values of B in such a way that the operations within A also map to corresponding operations in B, obeying the rules there. Such a "structure preserving" map is called a *homomorphism*.

For example, consider the algebra of Complex Numbers from Chapter 11 of The Book, represented as ordered pairs of real numbers. The projection map $\mathcal{R}e$([x,y]) → x preserves addition, but not multiplication: [w,x]+[y,z] = [w+y, x+z] and [w, x]*[y,z] = [w*y–x*z, w*z+x*y]. The left side of the first of these projects to w+y, as does the right side. But for the product case the left side projects to w*y and the right side to w*y–x*z.

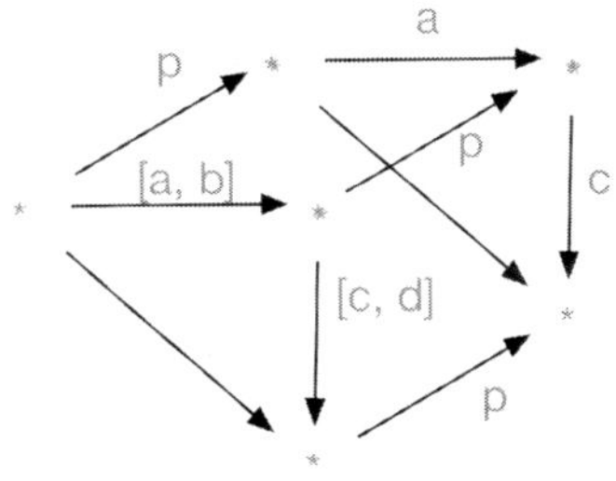

Here the "front" triangle is a Complex Number operation and P is its projection onto the real part. The unlabeled arrows (pointing down to the right) are the composed results of the operations. Consider what they would be if the operation is addition, vs what it is if it is multiplication. Using the diagram informally, in the first case we get a commutative prism, but not the other case.

If A and B are the same, a homomorphism is, more specifically, called an *endomorphism*. The identity mapping on any Algebraic Structure is structure preserving, so is an endomorphism.

Some homomorphisms are invertible. That is, for a homomorphism f:A → B there might also be a homomorphism g:B → A such that f∘g:B → A is the identity map on B and g∘f:A → B is the identity map on A. An invertible homomorphism is called an *isomorphism* and the two structures are somehow seen as two views of the same ideas. If there is an isomorphism between A and B, then A and B are called *isomorphic*. We then write A ≅ B. Isomorphism defines "sameness" in category theory: having the same information.

You can verify that there is an isomorphism between FACTORS(12) and FACTORS(20). And note that the projection function for the Complex Numbers onto the real part, even with just addition, isn't invertible. What can you say about FACTORS(12) and FACTORS(24)? How about FACTORS(12) and FACTORS(18). What is the same? What is different? Why is that so?

An endomorphism on A that is an isomorphism (invertible) is called an *automorphism*.

This terminology arose in Abstract Algebra, which covers, at least, Group Theory(+), Ring Theory(+), and Field Theory(+). It was noticed that identity and associativity are ubiquitous in those studies which led to category theory and its terminology - morphisms. Those ideas, in turn, applied to Topology(+) and Geometry(+), led to Algebraic Topology(+) and Algebraic Geometry(+).

It is important to keep in mind everywhere here that there are three basic elements in our study. The first is *data* as represented by, for example, the values in a set like the Integers. The second is the *structure* of the whole, operations, say. The third element is the *rules* that govern the structure. In a category, applying all of this, letting arrows flow through the structure implies "information". But combinations of data in various ways is also data. The compositions that connect 2 to 12 in FACTORS(12), for example is a kind of data.

Isomorphic Objects in a Category

In some categories, some objects that seem to be different are indistinguishable in any real way. Such objects are called *isomorphic objects*.

Suppose we have two objects, a and b, in a category $\mathcal{C}$. Suppose that we have an arrow f:a $\rightarrow$ b from a to b and another g:b $\rightarrow$ a. Finally suppose that fg is I_b and gf is I_a. Then we say that a and b are *isomorphic* and the f and g arrows are isomorphisms.

Moreover, in this case, f is called the inverse of g, since fg is I_b and g is called the *inverse* of f since gf is I_a. Note that either of these could be true without the other being true so we need both to define an isomorphism.

Thus g is the inverse of f provided that the following diagram is commutative.

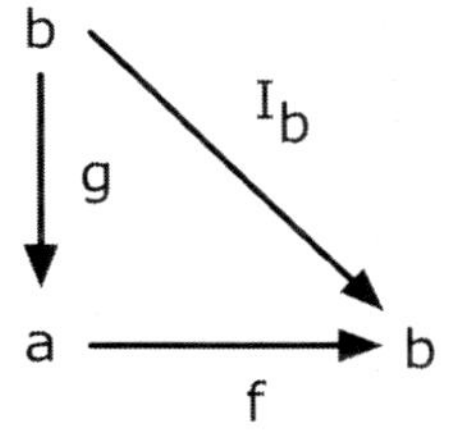

Likewise f is the inverse of g provided the following is commutative.

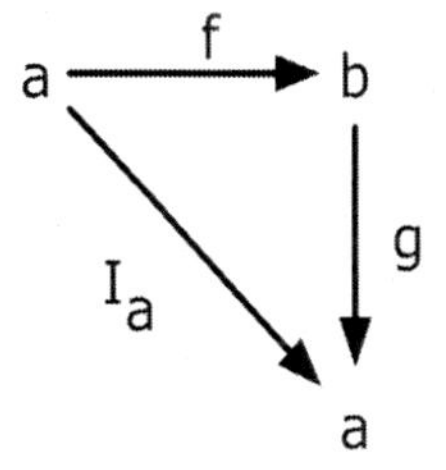

Finally a and b are isomorphic when the following forms a commutative square. Note that it just matches the g arrows of the above. We could also match the f arrows with an equivalent result. We will define isomorphic *arrows* in a category later; when two arrows are, in the category sense, "the same".

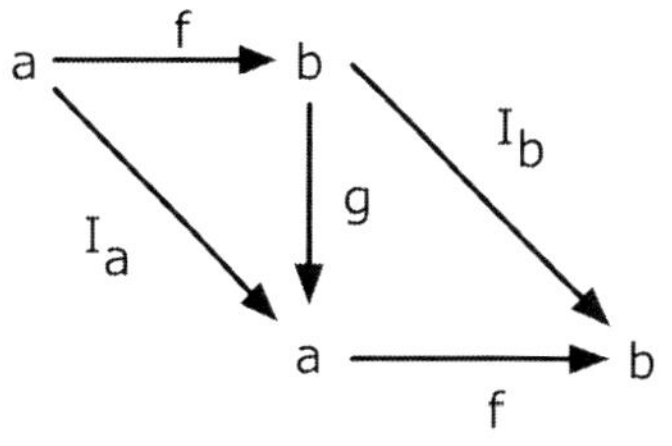

In the clock-like category with four objects, all of the objects are isomorphic.

In FACTORS(12) and similar categories, none of the objects are isomorphic to others.

2 Bridge – Structure

The Whole Numbers as a Category Using Addition

We can use the same ideas as we used in the CLOCK example to represent the Whole Numbers, {0, 1, 2,...}, as defined in The Book, as a category. We can do it in two ways, either as a monoid with $\mathfrak{W}$ as the single object or with the elements of the set of whole numbers as the objects. We will take the first approach here and let you explore the other.

Again, as in other monoids, the morphisms are the essential item, and an arrow will represent adding a value to "something" unspecified. This will work since adding 3, say, to five has the same effect (though not the same result) as adding 3 to 42. So, when we write x +3 we interpret x as the source object and +3 as the morphism. So, 5 + 3 = 8 looks like:

$5 \xrightarrow{+3} 8$, or, more properly, $\mathfrak{W} \xrightarrow{+3} \mathfrak{W}$

Again, in the monoid view, the Whole Numbers are not thought of as a set, but as an object with states. The morphisms represent the *values* as we normally think of them, but they can also be thought of as transitions between the various possible states of the Whole Numbers. So, in the left hand diagram just above, the 5 and the 8 are just *states* of $\mathfrak{W}$. It is the arrow that represents the value 3.

The arrow +0 serves as the identity element for this monoid and composition is just addition as defined in the Whole Numbers. That is to say, the set of arrows is isomorphic (the same as) the Whole Numbers. Associativity is just the associative property of addition since composition is just the operation on the values.

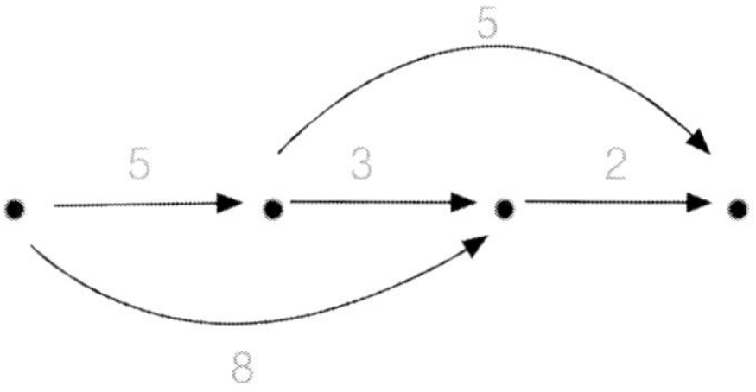

A more productive way to think about this, however, is that the morphisms are simply values of $\mathfrak{W}$ and composition of morphisms (arrows) is the addition operation. Then, composing 3 with 2 is the addition operation 3 + 2.

Note that this monoid has a property not required by categories in general, in that addition is *commutative*. So, following the +4 arrow with the +3 arrow has the same effect as applying them in the opposite order. You might want to draw the figures for this. Thus, category is an abstraction of the Whole Numbers since you remove a property, making Category a more general idea. This won't work for the Natural Numbers, however (why?), though you might want to think about the Natural Numbers and multiplication and see how the rules for a Category apply to that example.

If we consider the other (non-monoid) view of $\mathfrak{W}$ as a set with addition, then the Whole Numbers are objects and the arrows are functional transitions between them. In this representation, zero has a special place since there is a unique transition (morphism) from 0 to any other object. There are compositions that are equivalent to this morphism, but there is only one direct arrow from 0 to, say, 42. Such an object in a category is called an *initial object*. Note that FACTORS(12) and similar things have initial objects also. Likewise FACTORS(12) has a *final* object; one that has a unique arrow with that object as destination from every object in the category. Not all categories have initial or final objects. But note that an initial object in $\mathcal{C}$ is final in $\mathcal{C}^{OP}$.

If a category has an initial object it is unique (up to isomorphism) since two such have arrows to each other, but their composition would then need to be the identity arrow by the uniqueness property of initial objects. The same is true for final objects.

Contrasting these two views of the Whole Numbers we see that in the monoid view, the arrows (values) represent state changes with composition representing the addition operation, while in the non-monoid view they are compositions of two functions: One is the successor function, *succ*, for successor, as defined in The Book. The other is the identity function, *nop*, for no-operation. There is actually a relationship that connects and unifies these two views of this category (and others). See Arrow Categories, below. It is a deeper dive than we really need here, however.

We will see an alternate representation of the Whole Numbers below in the context of an arbitrary infinite set and functions defined on its subsets.

You can extend the ideas above to the Integers as a whole, of course, provided you consider the notion that negative numbers are just arrows that point in the opposite direction to the positive numbers, so that you have additive inverses and a group under addition. The Integers have no initial object under addition.

Also note that in computer programming integer operations are normally clock-like since a computer is a finite beast. Typically, though, there are 2^{32} or 2^{64} values on this "clock" and those values are offset so that some of the "hours" represent negative integers. Representing Integers in general is very difficult on a finite machine. Moreover, if you take types in the language as objects, then (pure) functions without side effects are morphisms on those types.

Functions From a Set to Itself

Let S be a set. Consider the set of total functions from S to itself.

A function, f, from set A to set B is called *total* provided that f(a) is defined for every *a* in A. That is, it isn't defined on just a subset of A. We assume all functions here are total unless we specify that they are *partial*; defined on a subset of the domain.

This collection of total functions, the monoid FUNCTIONS(S), has the identity function as the identity morphism. Functions compose properly and obey the associative law when composed. There is a single object and the morphisms are the functions. In very simple cases we can actually list and count the morphisms. If S is empty, then FUNCTIONS(S) is just SELF. If S = {0, 1, 2} we have nine such functions, since for any object in the domain of the function (3 possibilities) we have three possibilities for the value of the function. Notice that compositions of such functions actually yield one of these nine, also.

Note that you can start with a set of sets, instead of a single one. Those sets then become the objects of a category. The morphisms are all total functions from any one of those sets to any of them, including itself. We leave the details to the reader. You might want to experiment with the following three sets: { }=∅ (the empty set), {0}, and {0, 1}.

Compositions of a Set of Functions

Suppose we have a set S and a set F containing a few functions on S; total functions from S to S. Create a new set of functions F* that includes both the identity function on S and also all compositions of the original functions in F and any others in F*. Such a set is called *closed under composition* and also the *closure of F under composition*. For example, let S be the Whole Numbers and let F contains one function that adds a constant Whole Number value, such as f(x) = x+3. Suppose F also contains one similar multiplying function like g(x) = x*2. Then f* will contain both compositions f∘g and g∘f. The first of these is f∘g(x) = f(x*2) = x*2 + 3. The other is g∘f(x) = g(x+3) = (x+3)*2. Since these functions are in F* they can be composed with others, yielding even more complex results. We note that a function can be composed with itself, also.

Likewise g∘f(5) = (5+3)*2 = 16, mapping 5 to 16 which neither of the original functions did.

Suppose we have another such function, say, $h(x) = x*4$.

Then $h\circ(g\circ f)(x) = h((x+3)*2)) = ((x+3)*2)*4$

and, since $h\circ g(x) = h(x*2) = (x*2)*4$,

we get $(h\circ g)\circ f(x) = (h\circ g)(x+3) = ((x+3)*2)*4$,

illustrating transitivity even when we mix operators.

And, if S were some subset of the Whole Numbers, say the even numbers, then not all functions would work, since they would need to map evens to evens. The function f, above, wouldn't work, but g would.

Each such function "connects" an individual Whole Number to another; g(7) = 14, for example, connecting 7 to 14. So, interpreting differently, g *generates* an arrow from 7 to 14.

There is no arrow from 0 to either 1 or 2 in this structure, nor from 1 to 3. Think about the other "missing" arrows.

Note that in this interpretation, a function actually generates several arrows, not a single one. The function g "connects" 7 to 14 but also 3 to 6. So, treat the independent variable as a source and treat the result as the destination of an arrow. The identity function, since it is total, generates all of the identity morphisms if we treat these connections as arrows.

Then S with morphisms generated by functions in F* defines a category. The morphisms aren't the actual functions, but the connections that they generate, since an arrow (morphism) has both source and destination as essential components. Call this category COMPOSITIONS(F).

There is another way to look at this idea. The above doesn't define a monoid, assuming F and S are non trivial, But, since all the functions in F* are mappings of S to S we can consider S with morphisms F* to be a monoid. F* contains the identity (the identity function). All of its members compose and function composition is associative. Call this monoid CLOSURE(S, F).

Next, lets do the same, but imagine a much larger starting set F of functions from $\mathfrak{W}$ to $\mathfrak{W}$. There will be two kinds of functions. Let $f_n(x) = x+n$, for each Whole Number n. f_0 is the identity function, of course. Let $g_n(x) = x*n$, for each Whole Number n. g_1 is, again, the identity function in the Whole Numbers. Now take the closure of F and think about all we've seen above.

Time Out: Thinking About Things

Lets summarize a few things before going on. First, while we are only relating categories to numbers for the most part, note that they have applications far beyond such simple things. We have seen a bit about sets, of course, and will see more. But categories have wide application in higher math as well and form a unifying structure for many things.

The essence of a category is *identity*, *composition*, and *associativity*. These relatively simple things appear over and over in mathematics. Adding features to this basis gives new structures such as groups, for example. Likewise deleting features from complex things, while maintaining the basic category structure can lead to other things. Keep in mind, then, that we are only taking a dip into a very deep pond. For a much deeper dive, read Eugenia Cheng's book, *The Joy of Abstraction*, which expands on all of the ideas here.

Next, the *concept* of a morphism varies widely between different categories. In some, such as SUBSET(S) a morphism between subset a and subset b is just a true-false relationship. Set a is a subset of b or it is not. But in the Integers (which is a group, hence a monoid) the +k arrows are values that can be composed with other values according to an operation. The reader should verify the monoid nature of the Integers with the plus operation defining composition of morphisms. The plus sign (+), as used here has two meanings (called *overloading*). First it is part of the name of a function

(+k):x → x+k. It is also the name of the composition operator. Note that associativity of morphisms is essentially trivial in SUBSET(S) but it takes work to show it in the Integers. In general, the morphisms don't really need a "meaning" at all.

Any *directed graph*[+] with the addition of identity arrows is a category. (Why?) But you need to be a bit careful here if the graph has cycles. For example, consider a graph with two vertices and an arrow from each to the other. Say f:a → b and g:b → a. We get different categories depending on how we interpret f∘g and g∘f. Are they the respective identity morphisms or not? If not then we could actually have infinitely many arrows built from the compositions, sort of like a ping pong game or a blinking light. It is still a category, but not the same category as if at least one of the compositions is the identity.

There are some things that are "almost" categories, but not quite. If we take the Integers as objects with morphisms defined only by the less than operator, we fail to make a category since there are no identity arrows. That is, a < a is never the case. If we add identity arrows to this we end up with two kinds of arrows, "=", and "<" and the Integers with these morphisms is, indeed, a monoid. It is, however, different from the monoid of the Integers with "≤" defining the arrows since an "=" morphism carries more information than the "≤" one, as does a "<" morphism.

And, as with the FACTORS example, above, we need to interpret the common notation of < and such as arrows, not truth values. So a < b is an arrow from a to b with result (destination) b, permitting such things as (a < b) < c to make sense.

In a wildly different context, if we let the objects be the set of humans, then "child-of" doesn't define proper arrows both for failure of identity and composition. If a is a child of b and b is a child of c then a isn't a child of c, but a grandchild. Replacing "child" with "descendant" might work if you agree that people are a descendants of themselves. Note that the child - parent graph can simply be reinterpreted as a child - ancestor graph.

Still More Thinking: Semantics

Here is an issue that you can think about. Suppose we are working in some category $\mathcal{C}$ and we have a morphism, f, from a to b and another, g, from b to c. Suppose that we compose them, g∘f taking a to c. The semantic question is whether this composition *is* a morphism from a to c or only *guarantees the existence* of such a morphism. In other words is g∘f a member of $\mathcal{C}$(a, c). And note that neither f nor g is a member of that collection. Another way to think about it is in terms of composing those collections themselves. Can we think of $\mathcal{C}$(a, b) composed with $\mathcal{C}$(b, c), yielding $\mathcal{C}$(a, c)?

If you want to answer "no" to the first question, above, then note that in a directed graph there might not be an arrow from a to c other than the composition of two arrows through, say, b.

Suppose, in addition, that we know that $\mathcal{C}$(a, c) contains another morphism h, say, one that didn't arise from f or g. The question that then arises is whether g∘f *is the same* as h in some sense or potentially not at all. Can we write g∘f = h and have it meaningful? That is, is the source, arrow, and destination enough to declare equality?

And if you want to answer "yes" to this question then what happens in a directed graph with more than one arrow between a pair of vertices? The *free category*(+) on a graph has vertices as objects and all paths as arrows.

The Category SET(S)

Let S be a set. Consider the subsets of S as the objects of a category as follows.

The arrows of this new category, the morphisms, is the set of total functions from some subset of S to another. Note that all such functions are also functions to S itself, though we want to restrict the range set. Without such a restriction, the only composable functions would need to be total on S, which is more restrictive than we want.

The identity morphism for set A is the identity function on A: $I_A(a) = a$ for every a in A.

Composition in this category is function composition. If f is a total function from A to B and g is a total function from B to C then $g \circ f(a) = g(f(a))$ defines a total function from A to C. The identity and associativity rules are easy to verify for this category, called SET(S), since we start with a given universal set S.

Note the huge number of objects and morphisms here if S is non-trivial. But think about SET(∅), or SET({0}) defined over a singleton set, {0}.

In math, one doesn't often encounter a function from the empty set to some other, but we need it here. It is simply an empty set of pairs of values, using the ordered-pair interpretation of a function. Call it the *empty function.*

The initial object of SET(S) would be the empty set. The empty function is a morphism to any subset of S and the empty function is total only for the empty set.

If we extend beyond sets, there is a much larger category, like the above, but doesn't require that we start with a fixed set. That is beyond our purposes here: Category SET[+].

Applied to category theory, a morphism, f, from A to B is *invertible,* provided that there is another morphism, g, from B to A such that $f \circ g$ is the identity on B and $g \circ f$ is the identity on A. In this case, objects A and B are isomorphic, which is the notion of "sameness" in categories. In the category SET(S), those functions that are 1-1 and onto (injective and surjective) are invertible and so subsets of the same cardinality are isomorphic in this category. You can also examine variations of this idea. Suppose you restrict the functions to be surjective. Is that property preserved in composition? What happens to the empty function there?

Subcategories

A category $\mathscr{D}$ is a subcategory of a category $\mathscr{C}$ provided that the objects of $\mathscr{D}$ are also objects of $\mathscr{C}$, and that the Hom sets in $\mathscr{D}$ are contained in the Hom sets of $\mathscr{C}$. Actually, it is a bit more restrained. The identity arrows of $\mathscr{D}$ need to also be identities in $\mathscr{C}$, and the arrows need to connect in the same way in both categories. In other words, you can omit certain objects, but there are constraints on which homomorphisms can be omitted to form a subcategory. In particular, you need to keep identity morphisms for the objects in $\mathscr{D}$. Nor can you drop B, but keep the arrow that connects A to B, which wouldn't make any sense. A subcategory needs to remain a category.

For example, you get a subcategory of SUBSETS(S) by taking all finite subsets of S as the objects and keeping the subset relationship as the morphisms.

Similarly, you can form a subcategory of SET(S) by taking all of the finite subsets as objects and all of the functions among them as morphisms. FINITE(S) is a name for this category.

Another, simpler, example is to drop 3 from FACTORS(12) along with the arrows in and out of it to get a subcategory. Alternatively you could keep 3, but drop the arrow from 3 to 6, though the interpretation might not mean anything. You could not, however drop that arrow but add an arrow from 4 to 6. It would still be a category, but not a subcategory.

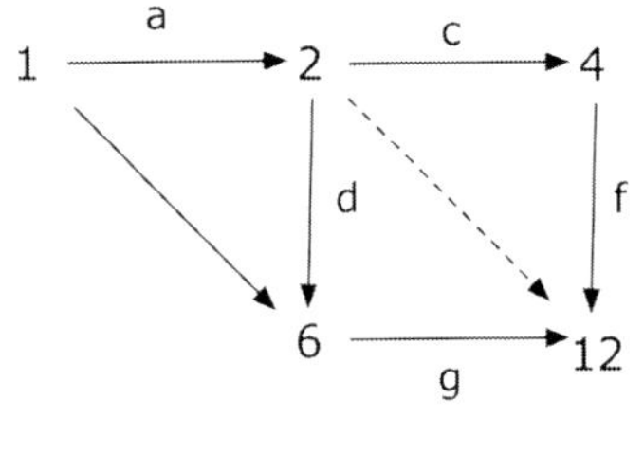

You can keep objects, but drop some arrows between them. In the next chapter we will see some other examples. Look for them.

Essential Elements and Natural Numbers Objects in SET(S)

And beyond...

Singleton subsets of S have a special place in SET(S). If A is any subset of S then there is a unique function from A to {*} for any singleton {*}. This includes the fact that there is a single function from the singleton to itself (the identity). This makes singletons into *terminal objects* and thus, all singleton sets, treated as objects, are isomorphic. But there are also functions from {*} to other subsets of S, including S itself. A morphism $e:\{*\} \to A$ *picks out* a single element of A, and such a morphism is called an essential element of SET(S). And, for a non-trivial set S, there are a lot of these. Distinguish *object* of a category and *essential element.* They aren't the same, of course. Essential elements, sometimes shortened to *element,* in general, are morphisms with a terminal object as source (domain). This notion extends to other categories that have terminal objects with morphisms to other objects.

Suppose that we have an essential element, $z:\{*\} \to N$, in SET(S) along with a morphism $s:N \to N$. Note that N is a subset of S, here, an object in SET(S), and s is a function. We can compose s with itself ss, sss, ssss, ... indefinitely. Let's look at two cases. Let n compositions of s with itself be denoted s^n.

(a) Suppose that $s^n z(\{*\}) \neq s^k z(\{*\})$ whenever $n \neq k$. This implies that N be infinite, of course and that, starting with $z(\{*\})$ we have something equivalent to the Natural Numbers, with $z(\{*\})$ taking the place of 1 (alternatively, 0) and s acting like the successor function. The pair of morphisms {z, s} is (perhaps) an example of a *natural numbers object* (NNO) in SET(S) and the concept generalizes to other categories with terminal objects. The full definition of an NNO appears later here and has some restrictions that must be met.

We can perform addition in this case also. Note that $sI_N = I_N s = s$, where I_N is the identity on N. Then 2 + 3 can be expressed as $sssI_N ssz(\{*\}) = ssssI_N sz(\{*\}) = sssssI_N z(\{*\}) = sssssz(\{*\})$. This last represents 5 if we take $z(\{*\})$ to be 0 (see NOTE). Moreover, we can express the commutativity of addition here since, for example, $sssI_N ssz(\{*\}) = ssI_N sssz(\{*\})$ by the same reasoning.

(b) Suppose that $s^n z(\{*\}) = s^k z(\{*\})$ for some n and k. If k is 0, so that $s^n z(\{*\}) = z(\{*\})$, then the result is a clock-like structure. This isn't an NNO for reasons that we explore below.

NOTE: And, we could just as well think of $z(\{*\})$ as representing 0 instead of 1, yielding a representation of the Whole Numbers. This is actually more "natural" and one of the reasons that modern authors express the Peano Axioms as "starting" at 0 rather than 1, which we used in The Book, keeping to tradition. So, recent usage has the *Natural Numbers* as {0, 1,...}.

The general definition of a Natural Numbers Object is a bit more complex, but this will do for here.

$$\mathfrak{N} \longrightarrow \mathfrak{W} \longrightarrow \mathfrak{Z} \longrightarrow \mathfrak{R} \longrightarrow \mathfrak{C}$$

3 Number and More

Whole Numbers With Addition and Multiplication

Thus far we have only considered addition arrows in the Whole Numbers. But we can also look at including multiplication as well and see if we still have a category. (Yes! Yes, we do!)

It is relatively easy to show that $\mathfrak{W}$ with multiplication as the operator, 1 as the identity, and the values as arrows is a category, with composition defined by the operator, but we want to consider the case of two separate operations.

Recall that multiplication in the Whole Numbers was defined using induction based on addition. In particular a * 0 = 0, a * 1 = a, and a * b′ = a * b + a.

Here we will take a monoid view, treating $\mathfrak{W}$ as the single object. We have to be careful, though, about the morphisms. Treating the values, alone, as arrows won't quite work. If we compose 3, 4, and 5 thinking of the first composition 3 and 4 to be addition and the second, 4 to 5 as multiplication, then it isn't associative. Grouped one way this is (3+4)*5 but the other it is 3+(4*5) and one of these things is not like the other.

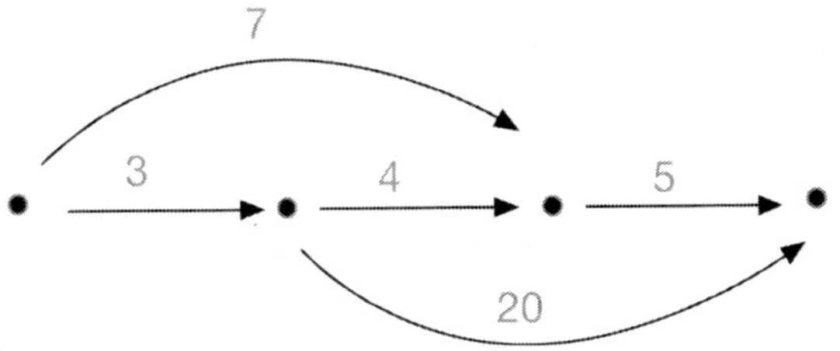

This is due to the fact that the 4 is used two different ways here. So, we need to be more sophisticated.

Instead of the values, we will use functions as the morphisms, and there will be a lot of functions; two families of functions parameterized by the Whole Numbers. We will use ideas from the above section on Sets of Functions.

For each Whole Number n, let $S_n: \mathfrak{W} \rightarrow \mathfrak{W}$ be defined by $S_n(x) = x+n$. We have infinitely many functions of this form. And S_0 is the identity function, of course. These functions can be composed:

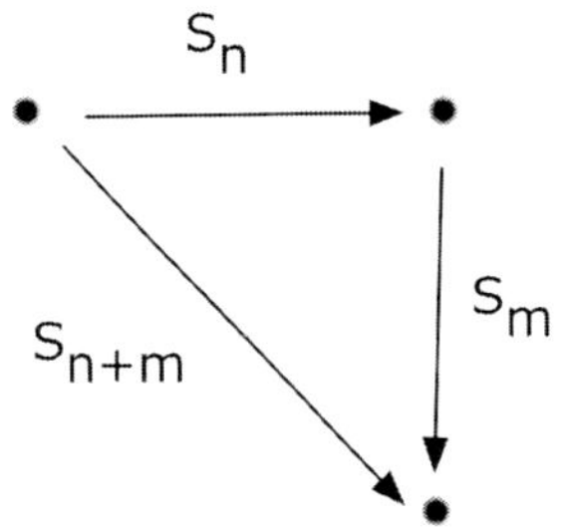

For each Whole Number n, let $P_n: \mathfrak{W} \rightarrow \mathfrak{W}$ be defined by $P_n(x) = x*n$. And, P_1 is, again, the identity function in the Whole Numbers. These can be composed similarly to the above, and we can also compose S and P functions.

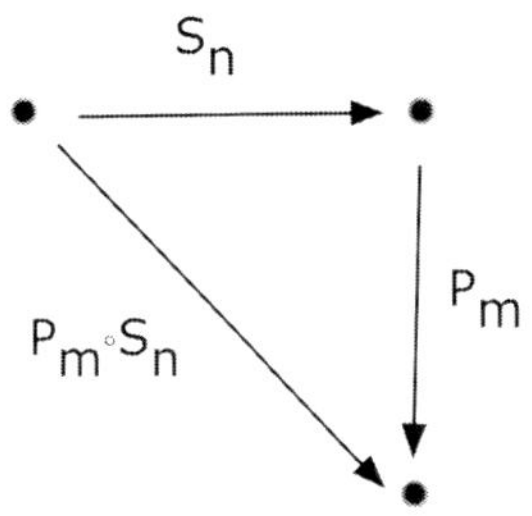

All of these are ordinary functions from $\mathfrak{W}$ to $\mathfrak{W}$, making them suitable as morphisms. Composition is function composition among these functions.

To verify associativity requires eight separate arguments, for the possible orderings of the two operations in groups of three. We have SSS, SSP, SPS, PSS, PPS, PSP, SPP, and PPP.

We will verify only one, say PSP. Consider the composition of P_k, S_m, and P_n using function composition.

Note that $S_m \circ P_k(y) = S_m(y*k) = (y*k)+m$.

Likewise, $P_n \circ S_m(z) = P_n(z + m) = (z+m)*n$.

So $P_n(S_m \circ P_k)(x) = P_n((x*k)+m) = ((x*k)+m)*n$,

and $(P_n \circ S_m) \circ P_k(x) = P_n \circ S_m(x*k) = (x*k+m)*n$,

and these both simplify to $x*k*n + m*n$. The other seven orderings are similar. Try a few.

The ideas of this section will be used over and over in what follows; functions as morphisms.

Note that both addition and multiplication in the Whole Numbers are commutative operations. For addition, this can be expressed in the following diagram:

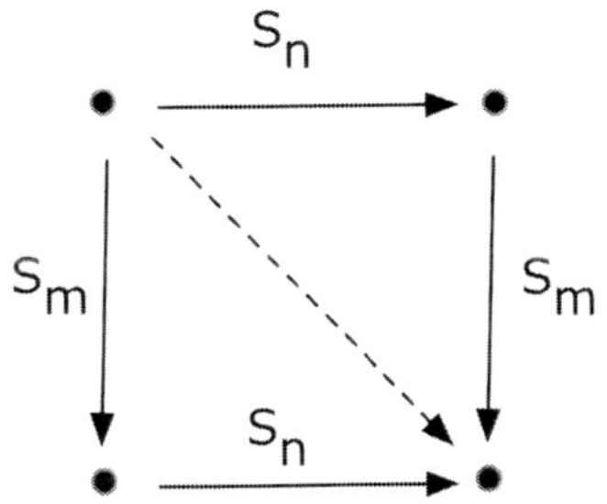

The statement that addition is commutative is the statement that this diagram is a commutative square. If we first add n and then m we get the same result as first adding m and then n. That isn't a proof, of course. And it wouldn't be true for any non-commutative operation. But a monoid is "commutative" if its operation makes this diagram commute. You are then in a position to consider functors (structure preserving mappings between categories, as discussed below) that preserve this commutativity.

Note that the mapping from n to S_n is injective (1-1), but not surjective (onto). The same is true about the mapping from n to P_n.

Alternate View of the Whole Numbers as a NNO

We assume we are working in SET(S) for some infinite set S and we choose some singleton set {*}, a terminal object. In the diagram below z and q are morphisms with the terminal object as domain (source). Those morphisms are called "essential elements", sometimes "global elements", or simply *elements*. Note that element and object are not the same concept. In particular an element is a morphism of a certain kind.

Then (definition) the pair of morphisms, {z, s}, is a *Natural Numbers Object* (NNO) provided that for *every* essential element q:{*} → A and *every* morphism f:A → A, there is a *unique* morphism u:N → A which makes the following diagram commute, both the triangle and the square. Note here that N and A are subsets of S. I've italicized some words in the above, since they are essential to the understanding below.

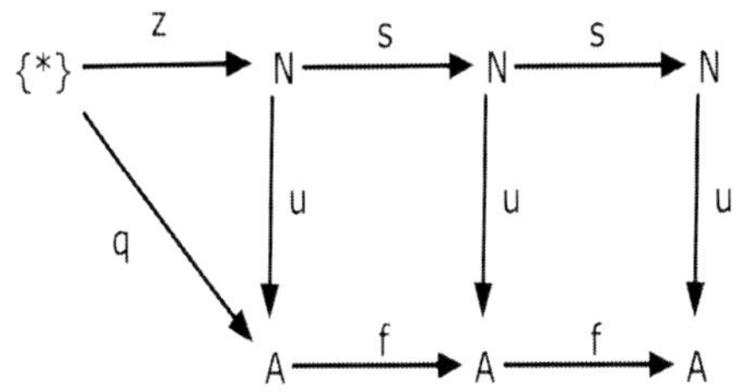

We don't emphasize it with the notation {z, s} for a NNO, but note that both {*} and N are part of this formalism, since morphisms include both source and destination.

Another way to state the above is to say that {z, s} *accommodates* the pair {q, f} if and only if there is a unique morphism u making the above diagram commute. Then {z, s} is a NNO if and only if it accommodates every such pair. We shall see that this puts restrictions on N and what it can and must contain. For some such z and s, there are functions that can't be accommodated.

Below, we explore the implications of this definition and will conclude at the end that it requires N be infinite and that it consist exactly of those elements of S that can be "reached" via compositions with z and s. N can consist of nothing less, and nothing more. Therefore, since it has a start point, it is just like what we have defined for the Whole Numbers. The argument is quite long.

Recall that z({*}) is an element of set N, and so are sz({*}), ssz({*}), etc. Let s^n represent composing n applications of function s.

First, consider $s^n z(\{*\})$, an element of N. Suppose, first, that $s^n z(\{*\}) \neq s^k z(\{*\})$ if $n \neq k$. We include the case k = 0; $s^0 z(\{*\}) = z(\{*\})$ here, so $s^n z(\{*\}) \neq s^0 z(\{*\}) = z(\{*\})$ if $n \neq 0$. This is only possible if N is infinite, of course, but we shall see that the definition of NNO puts an additional restriction on N. Not every infinite set, N, will do (*uniqueness* of u (given q and f) is the key). If N is *too big* then not every q and f can be accommodated.

Now, suppose that {z, s} is, indeed, a NNO. Some examples to illustrate the meaning of this follow.

Note that A cannot be empty because q is an essential element. But it need not be infinite since the definition says "every essential element q and function f".

Suppose A = {bird, cat, dog, elephant}, a finite set. Also note that q({*}) is some particular element of A, say, *bird*. And f is a total function on A and so defined for every element of A. There are many possibilities for f that need consideration including the identity function, constant functions, clock-like functions, etc. For each possibility we need a unique morphism u.

For any given function f, let u be the unique morphism (up to isomorphism) that makes the above commute. Then, thinking of z({*}) as being 0, we must have uz({*}) = "bird", also, to make the triangle commute. That is independent of f.

We consider a few possibilities for $f{:}A \rightarrow A$. Suppose, first, that it is a constant function with range set {*dog*}. Then, for any $t \in N$, including t = z({*}), fu(t) must be *dog*, also, and thus, us(t), must result in *dog* for the above square to commute. Also, note that compositions of f with itself will always end up at *dog* by the assumption f is a constant function.

However, the following diagram must also commute as a composition. So uss(t), ending in the bottom right corner, must also be *dog since f is applied twice to get there.*

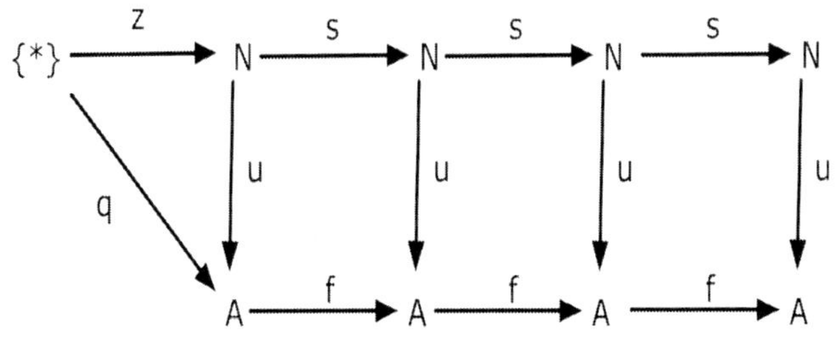

To simplify the notation here lets denote z({*}) as 0, sz({*}) as 1, ssz({*}) as 2, etc. This simplifies the expression of the matchings of u; namely u(0) = *bird* and u(t) = *dog* for other values of t, though we need induction to show it, extending the above diagram to the right.

But this says nothing about those elements of N that s can't "reach" through composition with itself from z({*}). Consider the following:

Suppose there is a value, y, in N such that $s^n z(\{*\}) \neq y$ for every n. Then u(y) could be anything and still have a commutative (extended) diagram. It would seem that u could map such elements anywhere meaning there are several such functions making the diagram commute, contradicting uniqueness. In other words, if N contains such *unreachable* elements as y then there are pairs {q, f} that can't be accommodated because of the uniqueness requirement on u.

We are forced, then to consider something like *surjections*, only, for compositions of function s. Thus, in an NNO, there is a restriction on N that it consist of *only* elements that can be reached by compositions of s starting at z({*}) like the above. So, an arbitrary infinite set for N and function s might not work, unless s "reaches" all of it through composition. Thus, the restriction of uniqueness implies that the set N in {z, s} is "exactly" like the Whole Numbers (Natural Numbers), rather than a larger set that (properly) contains something like the Whole Numbers. This, for me, is a major insight.

So, N is "nothing more" than the Whole Numbers. We will explore, in a bit, whether it can be less. Note we've forced N to be infinite in our original assumptions, of course. We still need to show that it *must* be infinite.

A more interesting case than a constant function is when f is, instead, the clock-like transformation of A that we saw earlier. Then u(0) = uz({*}) must be *bird*, u(1) = usz({*}) = fuz({*}) = *cat*, (similarly) u(2) = *dog*, u(3) = *elephant*, u(4) = cat, repeating the sequence. Thus, our NNO seems to be "counting" mod 4 in A.

Yet another case, is one in which {q, f} also forms an NNO. Then u is, in essence a morphism in which s "counts" A, so that three applications of s is the same as three applications of f. A could be larger, however. We can *count* in the Reals.

Something similar will need to happen if the image of f in A is finite even if A is infinite.

Next, we show that N *must* be infinite so we can drop the assumption and that the sequence of applications of S to z({*}) can't loop back, resulting in a finite set of elements of S. If N is finite then some pairs can't be accommodated.

We look at a case in which the s morphism has a "finite expression", so that s^nz({*}) = z({*}) for some n>0, though this is not the most general case but it should do. This will result in a contradiction to the statement that {z, s} is an NNO.

Here we are only considering what happens in N, not in A at the moment. The implication, again coming from a uniqueness argument is that N is also finite with n elements. Let's suppose that n = 2. Then ssz({*}) = z({*}). So, N = {z({*}), sz({*})}, with only two elements. As above, if there are more elements in N then we can't guarantee that any morphism u:N → A will be unique. We will look again at what u must do in such a case.

Assume that q:{*} → t, and that A has at least two elements. As before, what if f is a constant function onto a constant, x ≠ t? Then uz({*}) must be q({*}) = t, as before (triangle commutes) and usz({*}) will need to be x (square commutes and f is constant).

But now we seem to have a contradiction since x = fusz{{*}} = ussz{{*}} = uz{{*}} = t (both commute). And there is no guarantee that f actually has t = x. We conclude that a finite N isn't actually possible for an NNO. The first equality in the above is from the fact that f is constant so the application of f results in x. The second equality comes from commutativity of the diagram. The third equality is our assumption of the finiteness of N. The fourth is the fact that the triangle commutes. All but the third are possible and/or required, so the third must be false. So, this pair {q, A} can't be accommodated so {z, N} isn't an NNO.

The conclusion, then, is that an NNO implies an infinite set N and that {z, s} defines just the Whole Numbers, not a superset.

There is a further interesting case, however. Suppose that, starting with the essential element q we have the situation where compositions of f form a finite set. Staying with our {bird, ... elephant} example for set A, suppose that q({*}) = bird. Suppose also, that f is the function that maps each of the elements to the next as listed: f(*bird*) = *cat*, ... f(*dog*) = *elephant* and, cycling back, f(*elephant*) = *bird*. How will an NNO handle this case? Since the triangle commutes, the unique morphism u that makes everything, including extensions to the right commute, implies that uz({*}) = bird, usz({*}) = cat, ... usssz({*}) = elephant, and, also wrapping, ussssz({*}) = bird, again. Thus u is counting modulo 4 in this case.

The Integers As a Category (Addition and Multiplication)

RECAP: Recall that the Integers were defined using an equivalence relation on pairs of Whole Numbers. Two pairs (a, b) and (c, d) are equivalent if a + d = b + c using addition as defined in the Whole Numbers. The Integers are formally represented by the equivalence classes under this relation.

The Integers were defined to be equivalence classes of pairs of whole numbers, such as E(5, 3) = E(2, 0) = 2 and E(3, 5) = E(0, 2) = -2. All of the operations were defined in terms of operations of Whole Numbers acting on the components of representative pairs for these classes. This let us define and investigate something new in terms of things already known, which is a fundamental practice in mathematics.

Note that each such equivalence class, E(a, b) contains either a pair (n, 0) or (0, k) for Whole Numbers n or k. The former are the non-negative values and the latter are non-positive. Those pairs are considered "canonical" representatives of their respective classes. E(n, 0) is usually written n or +n in everyday math and E(0, n) is written -n.

Furthermore, addition was defined on these equivalence classes as E(a, b) +E(c, d) = E(a+c, b+d), or component-wise. Multiplication is a bit messier E(a, b)*E(c, d) = E(a*c+b*d, a*d+b*c), again using Whole Number addition and multiplication.

We hope to use ideas from a previous section to show that the Integers with a set of functions as arrows is also a category.

Define two classes of function S and P as follows.

Let $S_{[a,b]}$(E(c, d)) = E(a+c, b+d), where the subscript is a shorthand for an equivalence class E(a, b).

Similarly let $P_{[a,b]}$ (E(c, d)) = E(ac+bd, ad+bc).

Notice that these are functions on equivalence classes giving equivalence classes as results. Also, and importantly, note that a, b, c, and d here are Whole Numbers; {0, 1, 2, ...}.

Consider the Integers, $\mathbb{Z}$, as a monoid with these functions and their compositions as the morphisms. The identity morphism is $S_{[o,o]}$, or $P_{[1,0]}$, which are equivalent. Use similar, if messier, calculations as in the previous section to show that compositions of these morphisms are associative in any combinations.

It actually takes an argument to show that $S_{[o,o]}$, is equivalent to $P_{[1,0]}$, but the formal verification requires a look at Arrow Categories, which is to come. But if $S_{[o,o]}$ is an identity, we can demonstrate that it is isomorphic to $P_{[1,0]}$. Come back later and show the commutative diagram establishing this.

Also note that the examples in this and the previous section are really just specializations of SET(S) with different universes and restrictions on the functions (morphisms) that are considered.

Time Out: More Thinking

Stay aware of the fact that the operations on the Integers were all defined in terms of operations on the Whole Numbers. That was the purpose and benefit of the pair idea. We can take the known (Whole Number ideas) and use them to extend our range. This was a main theme of The Book, but we point it out here since it is a main driver of mathematical thinking. Thus, while the notation got a bit messy (too messy?), the new ideas have a firm foundation on things going back to the Peano Axioms.

In fact, category theory in general is a lot about formal manipulation of symbols and the chasing of connections. We are about to dive a bit deeper.

Functors, Embedding, and Relationships Between Categories

So far we have only discussed how some simple categories are an abstraction over some common mathematical structures. The concept of category captures common features of many mathematical systems. But you can also look at relationships between categories and how that can formalize relationships between other mathematical systems. First we will look at the relationship between the Whole Numbers and the Integers as developed in The Book.

A *functor* F is a mapping from a category $\mathscr{C}$ to a category $\mathscr{D}$ that preserves the structure of the category. This means that objects of $\mathscr{C}$ map to objects of $\mathscr{D}$, identity arrows of $\mathscr{C}$ map to identity arrows of $\mathscr{D}$, and the arrow relationships generally in $\mathscr{C}$ map to equivalent morphisms in $\mathscr{D}$. More formally:

> A functor F associates an object, b, in $\mathscr{D}$ with each object a in $\mathscr{C}$: $F(a) = b$ is an object of $\mathscr{D}$ $\forall a \in \mathscr{C}$, and
>
> $F(I_a) = I_{F(a)}$ for every identity morphism I_a in $\mathscr{C}$, and
>
> $F(g \circ f) = F(g) \circ F(f)$ for every pair of composable morphisms $f{:}a \to b$ and $g{:}b \to c$ in $\mathscr{C}$.

Since a functor preserves category structure, it is an example of a homomorphism.

To define a functor, F, from the Whole Numbers to the Integers we will use ideas from The Book (Chapter 6).

Here we will use the usual notation for Whole Numbers. We see how to "embed" the Whole Numbers into the Integers (using the functor), justifying our shorthand thinking that the Whole Numbers are "just" a subset of the Integers.

Define F as follows: For each Whole Number n, let F(n) = E(n, 0). That is, the result of applying F to a Whole Number n is the equivalence class that contains the pair (n, 0). Thinking informally this equivalence class is just the Integer n - 0, of course.

We also have to see how the arrows in the Whole Numbers map to arrows in the Integers, recalling that the morphisms were certain functions. So, if S_n and P_n are arrows in the Whole Numbers, let $F(S_n) = S_{[n,0]}$ and $F(P_n) = P_{[n,1]}$. The right hand sides of these are arrows in the Integers, of course.

Let's see how these work with commutative squares:

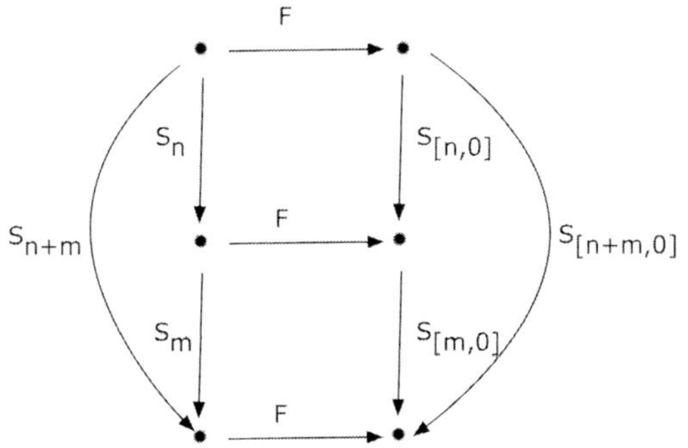

Something similar will be seen for the P morphisms. Yes, they commute. If we mix operations, the picture is messier, because we don't have a simplified notation for compositions in general. It looks like this:

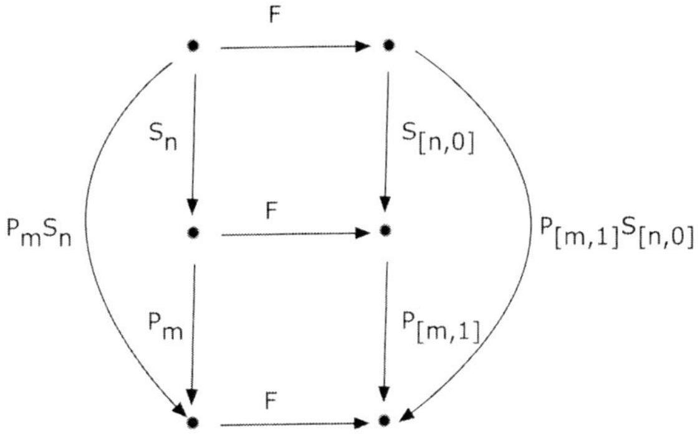

Note, that $P_mS_n(x) = P_m(x + n) = (x+n)*m$, and $P_{[m,1]}S_{[n,0]}([x, 0]) = P_{[m,1]}([x + n, 0]) = [(x+n)*m, 0]$, so we get a commutative diagram even here.

Think about embedding the Integers into the Rationals in the same way, as we will explore soon here. Or how the Rationals can be embedded into the Reals. Note that for a Rational number, r, the sequence $a_n = r$ for every n is a canonical representative of the equivalence class of r within the Real Numbers. Likewise the Real Numbers can be embedded into the Complex Numbers with a simple functor. That doesn't imply that all of these number systems are the same, since each in turn adds new structure. In other words, there may not be a reverse functor for any given one.

Note that we can't quite do the above for embedding the Natural Numbers into the Whole Numbers since the Natural Numbers aren't (quite) a category. That also helps explain, perhaps, why the development of the Whole Numbers by adding just zero to the Natural Numbers was ad hoc. The later developments feel much "cleaner" in some mathematical sense.

Moreover, the distributive law of multiplication over addition isn't represented here.

Hom Sets as Objects

Suppose we have a category $\mathcal{C}$. If a and b are objects in $\mathcal{C}$ then $\mathcal{C}$(a, b) is the set of morphisms from a to b. Suppose we want to consider these sets as objects in some new, higher level, category with functions between these sets as morphisms? Some of these sets are empty, of course, but $\mathcal{C}$(a, b), as a set, has at least the identity function to itself which will be the identity morphism on $\mathcal{C}$(a, b) considered as an object. Note that all of these sets are subsets of Hom($\mathcal{C}$).

Recall that SET(S) is a category with subsets of a set S as objects and all functions between subsets as morphisms. We will look for a sub category of this, since the Hom sets of pairs of objects of a category $\mathcal{C}$ are subsets of the collection (here a set) of morphisms, subsets of HOM($\mathcal{C}$). So, we are looking inside SET(HOM($\mathcal{C}$)), for some category $\mathcal{C}$. We will be looking for functions between these HOM sets; ordinary functions where the domain is one HOM set and the range is another. We require such functions to be total, as usual. There is more than one way to do this, but we shall only look at one of them based on the idea of a *pushforward*.

Before we start, note that in keeping with other things here we are looking only at small categories. That is, categories where the collection of objects is a set and the collection of morphisms is also "small" enough to be a proper set. So, our beginning category needs to be a set with only a set's worth of arrows. Category theory in general goes far beyond this restriction. In particular for such a category $\mathcal{C}$, $\mathcal{C}(a, b)$ is a, possibly empty, set for each pair, a and b, of objects of $\mathcal{C}$.

Notationally we will also drop the explicit punctuation mark indicating a composition and use simple juxtaposition in what follows to indicate composition. So, instead of $g \circ f$ to indicate the composition we will write, simply, gf. This is actually normal usage in category-speak.

So, let $\mathcal{C}$ be a (small) category, though what we say extends beyond sets. Consider any morphism $f{:}X \rightarrow Y$, where X and Y are objects of $\mathcal{C}$. Now, suppose that there is another (actually, one or more) morphisms that have X as the destination, say $g{:}Z \rightarrow X$. Then there is at least one morphism, fg, from Z to Y, There might be more. Note that g is in $\mathcal{C}(Z, X)$ and f is in $\mathcal{C}(X, Y)$ and fg is in $\mathcal{C}(Z, Y)$. Thus, if we think of mapping g to fg we have a connection between $\mathcal{C}(Z, X)$ and $\mathcal{C}(Z, Y)$.

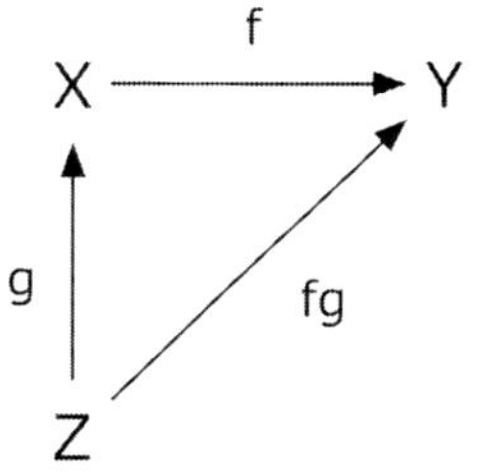

This connection is the start of a function between those latter two HOM sets.

$$\mathcal{C}(Z,X) \xrightarrow{f_*} \mathcal{C}(Z,Y)$$

It is a (total) function, however, since we can do exactly the same thing or any (and all) morphisms from Z to X. This function, f_* is called the pushforward of f, since it "extends forward" any "move" from Z to X to one from Z to Y. Consider, also, that this mapping actually a morphism from

$\mathcal{C}(Z, X)$ as the source to $\mathcal{C}(Z, Y)$ as the destination, considering the hom sets as objects. We state without proof that X and Y are isomorphic, with isomorphism f, if and only if f_* is bijective; that is if f_* is a set isomorphism. We shall return to morphisms of arrows in a future section. There is a corresponding notion of a pullback(+), as well.

Note that if the g morphism above is actually the identity on X, which takes X to X, then fg is just f itself so the mapping f_* from $g = I_X$ to $fg = fI_x = f$ implies that f_* is an identity in this case. Finally, this mapping from f to f_* when extended to all morphisms defines a functor.

So, to summarize: for any morphism $f{:}X \to Y$, and any Z in C, there is a function (hence a morphism) $f_*{:}C(Z, X) \to C(Z, Y)$. Since the identity function on sets is an identity morphism, and since function composition is associative, this makes our SET(HOM(C)) into a category. Ah, but we need an example, though it will be pretty extensive, since even small categories can have a lot of morphisms.

Let's look again at FACTORS(12) which is big enough to demonstrate composition of the pushforward arrows, but not quite big enough to demonstrate associativity. Here, again is FACTORS(12) with the arrows labelled. We will need these arrow names to discuss sets of them, of course.

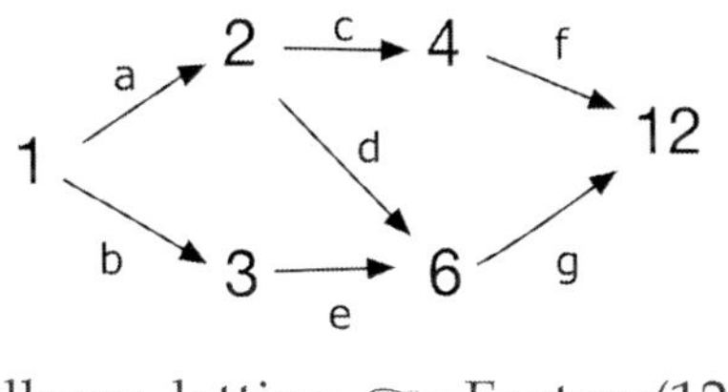

We list all the HOM sets as follows, letting $\mathcal{F}$ = Factors(12):

$\mathcal{F}(1,2) = \{a\}$, $\mathcal{F}(1,3) = \{b\}$, $\mathcal{F}(1,4) = \{ca\}$, $\mathcal{F}(1,6) = \{eb, da\}$, $\mathcal{F}(1,12) = \{fca, gda, geb\}$,

$\mathcal{F}(2,4) = \{c\}$, $\mathcal{F}(2,6) = \{d\}$, $\mathcal{F}(2,12) = \{fc, gd\}$,

$\mathcal{F}(3,6) = \{e\}$, $\mathcal{F}(3,12) = \{ge\}$,

$\mathcal{F}(4,12) = \{f\}$,

$\mathcal{F}(6,12) = \{g\}$.

These twelve sets will be the objects in our new "pushforward" category. Next, we have to look at some morphisms between these and it may not be immediately obvious which functions we should choose. Note that HOM(FACTORS(12)) is the union of all of the above sets along with the required identities, not just the immediate arrows a...g.

Caveat: We will see later that there aren't quite as many morphisms as we listed above, since some of them are isomorphic to each other, though we haven't yet defined what we mean by isomorphic arrows.

Let's start with arrow e in FACTORS(12) and compute it's pushforward. Note $e{:}3 \rightarrow 6$, with those numbers indicating objects. There is only one arrow with 3 as destination, namely b. And the composition eb is therefore a morphism from 1 to 6. It is the only such morphism, actually, since b is the only morphism from 1 to 3. From the above $\mathscr{A}(1,3) = \{b\}$ and $\mathscr{A}(1,6) = \{eb, da\}$, so matching b to eb is a total function between these HOM sets. That is $e_*{:}\{b\} \rightarrow \{eb, da\}$ defined by $e_*(b) = eb$ is a total function. That gets us started. To see composition, we take another starting point and see what develops.

Consider arrow $g{:}6 \rightarrow 12$ in FACTORS(12). We will compute g_* and see that it can be composed (as an ordinary function with e_* from the above. We look at how object 1 maps to 6: $\mathscr{A}(1, 6) = \{da, eb\}$. Both of these compose with g, giving gda and geb, respectively. The latter two composition morphisms are both members of $\mathscr{A}(1, 12)$, so that composition with the two elements actually defines a function, once again, from $\mathscr{A}(1, 6)$ into $\mathscr{A}(1, 12)$. That is $\{eb, da\}$ maps into $\{fca, gda, geb\}$. Namely, $g_*\,(eb) = geb$ and $g_*\,(da) = gda$. Do you see the pattern? This mapping is g_*.

However, we can also compose the functions defined in the two examples as g_*e_* mapping $\mathscr{A}(1, 3) = \{b\}$ into $\{da, eb\}$ and then into $\{ca, gda, geb\} = \mathscr{A}(1, 12)$.

Note, however, that the pushforwards of an arrow aren't unique for that arrow. We started with, not only an arrow, like g, but also with a particular arrow, eb, that mapped 1 to 6. We could have started with da instead.

Time Out: What's missing?

We've now seen two instances where we can use functions to represent operations in number systems and we can compose the simple functions S and P to get more complex functions. But we can't get everything. In particular, we can't represent $f(x) = x^2$ as a composition of the S and P functions. You might want to try it. The basic reason is that those functions are parametrized by constant values.

But, what we can do is add more functions to the basic set of morphisms, provided that we maintain the associative rules on compositions. For example, we could add a new class of exponential operators, noting that they are different depending on whether the independent variable is a Whole Number or an Integer:

Let $E_n(x) = x^n$ for any Whole Number n. (Even in the Integers we want that restriction.)

We need to be careful, however, and define 0^0 as 1, so that the functions (morphism) are transformations on the monoid. It won't do to leave it undefined.

Then, all of the S, P, and E functions are composable and maintain associativity. We still can't get $g(x) = x^2+2*x+1$, however, but we can get $h(x) = (x+1)*(x+1) = E_2 \circ S_1(x)$. That is interesting, but we don't, yet, have the distributive law represented.

Exponentials can also be included in what follows, but we will omit the details. Constant functions could also be added, but we will have no use for those.

The Rational Numbers As a Category

RECAP: In The Book (Chapter 7) we define the Rational Numbers as equivalence classes of pairs of values where the first component is an Integer

and the second is a non-zero Integer. Two pairs (a, b) and (c, d) are equivalent under this relation, which we will denote β, provided that a*d = b*c using multiplication in the Integers. An equivalence class, hence a rational, looks like β(-3, 4), for example, representing the value normally written as -3/4. In the following, the equivalence class notation will be useful on occasion. Canonical representations of a class are pairs in which the components have no common factors and the second is positive. So, the canonical representation of the class that contains (20, -6) is (-10, 3). Addition for pairs representing a Rational Number is defined as (a, b) + (c, d) = (a*d+c*b, b*d), using Integer operations within the parentheses. Note that neither b nor d is zero, so neither is their product. Of course, this result is only a representative of a certain equivalence class.
A pair equivalent to (n, 1), for an Integer n, represents an Integer within the Rational Numbers according to the functor we define below. That is to say, that such values behave just as the Integers do as well as being the image of the Integers within this new category.

We proceed as above, defining a collection of functions representing the operations we are currently interested in, addition and multiplication. Note that the functions are mappings of equivalence classes and the values a and b used to represent a class β(a, b) are Integers with $b \neq 0$. In the following, we will use a/b to represent this class for simplicity of notation, but also because it is how we normally represent rational numbers. Here we will name the function classes S′ and P′ to distinguish them from the functions used earlier.

Let $S'_{a/b}(x/y) = (a*y+b*x)/(b*y)$.

More formally this is

$S'_{\beta(a/b)}(\beta(x, y)) = \beta(a*y+b*x, b*y)$, but we prefer the more compact notation.

Likewise $P'_{a/b}(x/y) = (a*x)/(b*y)$.

Then $S'_{0/1}$ serves as an identity object, as does $P'_{1/1}$.

As above, these functions are composable, and the compositions obey the associative rule required so that the Rational Numbers with these morphisms form a category.

As an illustration, here is the verification that $S'_{0/1}$ serves as an identity both for left and right compositions. The left composition of $S'_{0/1}$ with, for example, a P′ morphism looks like

$S'_{0/1} \circ P'_{a/b}(x/y) = S'_{0/1}((a*x)/(b*y)) = ((0*b*y + 1*a*x) /1*b*y) = (a*x/b*y) = P'_{a/b}(x/y)$.

The right composition is

$P'_{a/b} \circ S_{0/1}(x/y) = P'_{a/b}((0*y+1*x)/(1*y)) = P'_{a/b}(x/y)$.

There are several other required computations to complete the verification, both of identity and association. But they follow a pattern like the above. Try a few.

Moreover we can embed the Integers into the Rational Numbers with a simple functor G. Let G(n) = β(n, 1) for any Integer n. Let the functor map an addition morphism in the Integers, say, $S_{[a, b]}$ to the morphism $S'_{(a-b)/1}$) for any Integer n = E(a, b). In other words adding n in the Integers adds n/1 in the Rational Numbers. Consider that the following diagram is commutative. In the right column, the operations within the pairs are Integer operations. We can do the same with the P and P′ morphisms.

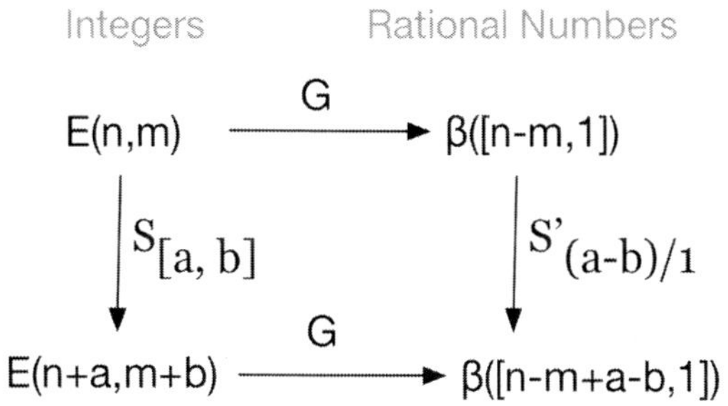

G will perform similarly with P′ morphisms as you can verify.

The Real Numbers As a Category

RECAP: In The Book (Chapter 9) we define the Real Numbers as the set of equivalence classes of Cauchy Sequences of Rational Numbers where sequences $\{a_n\}$ and $\{b_n\}$ are equivalent provided that the sequence $\{a_n - b_n\}$ converges to zero. We will denote this equivalence relation by ψ when needed. Note that a constant sequence of Rational Numbers converges to that same value. In particular, the sequence $\{b_n \mid \forall n, b_n = r\}$ will be denoted $\{r\}$ for any Rational Number r. The equivalence class containing $\{r\}$ is the number r, itself, interpreted as a Real Number: $\psi(\{r\}) = r$.

Note that the sum and product of Cauchy Sequences are both defined "term-wise" by taking the sequence that is, respectively, the sum or product of the individual terms of the two sequences.

We will use a notational shorthand for an equivalence class. What is formally $\psi(\{a_n \mid n = 0, 1, 2, \ldots\})$, we will write $\psi(a_n)$, omitting the setbuilder notation. For a constant rational sequence we omit the subscript n. So $\psi(\{1\})$ will become $\psi(1)$, the equivalence class of the constant sequence of all 1's. With this shorthand we define two collections of functions, similar to what we have done several times already.

Let $S''_{\psi(an)}(\psi(b_n)) = \psi(a_n + b_n)$, and

$P''_{\psi(an)}(\psi(b_n)) = \psi(a_n * b_n)$. That is, term-wise addition and multiplication.

With these morphisms, The Real Numbers are a category, that includes, in its own, quirky way, both addition and multiplication. The identity is either (or both of) $S''_{\psi(0)}$ or $P''_{\psi(1)}$, as should be obvious, but needs proof.

There is a functor, H, embedding the Rational Numbers into the Real Numbers as indicated in the commutative diagram below, though again showing only the Sum transformations:

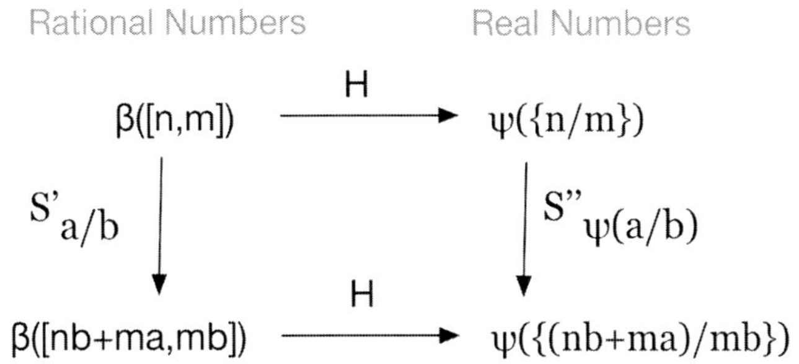

Real Polynomials as a Monoid

Consider the set of Real polynomials: polynomials in one variable with real coefficients and with the Real Numbers as the single independent variable. Treat them as arrows in a monoid, where composition of the arrows is either (or both) of multiplication and addition. This is a bit different from function composition, of course. The polynomial f(x) = 0 is an additive identity, and polynomial addition is associative. Likewise g(x) = 1 is a multiplicative identity and polynomial multiplication is associative. So, with either operation, Real polynomials form a monoid. But, as with the Whole Numbers, we will have trouble using both operations. Associativity won't work if we compose with addition and then with multiplication, for example.

Real Polynomials as a Category

To get a single category with both addition and multiplication we can, instead treat polynomials as objects and certain restricted functions (also polynomials) as arrows. This is the same idea used earlier for Whole Numbers.

Let Poly be the set of Real polynomials. In the following we consider only functions from Poly to Poly. So, if p is a polynomial let $S_p(q) = p+q$ for any Real Polynomial q, using polynomial addition. Likewise, let $P_p(q) = p*q$, using polynomial multiplication. Note that these two functions, S and P map functions to functions (polynomials to polynomials), not Reals to Reals. For example if $p = x^2+1$ then $S_p(3x+7) = x^2+3x+8$.

Once again, however, we need to verify associativity of these morphisms. Again, there are eight possible orderings of three items from a set of two. SPS is one of them. We need to show that S(PS) = (SP)S for any three such polynomial functions. $S_a(P_bS_c) = (S_aP_b)S_c$ for polynomials a, b, and c. The verifications are straightforward, though the distributive law for the Reals is needed to make the pieces fit together. Thus Poly is a monoid.

The Complex Numbers As a Category

RECAP: In The Book (Chapter 11) we define the Complex Numbers as a certain interpretation of pairs of Real Numbers with certain operations using the operations of the Real Numbers. In an ordered-pair (a, b), *a* is called the *real part* and *b* is called the *imaginary part*. The normal notation for a complex number is a + b*i* where *i* is taken to be the square root of -1. The operations are consistent with this view as $i^2 = (0, 1)^2 = (0, 1)*(0, 1) = (-1, 0)$. Note that this definition doesn't rely on equivalence classes. Two complex numbers are equal only if both the real parts and the imaginary parts are respectively equal. And note that the "parts" are both Real Numbers.

As with the Rational Numbers, students have a pair-like view of complex numbers even without the notation. And, unlike the earlier number systems here, this one isn't linear. It takes two dimensions to graph a Complex Number and four to graph a function from the Complex Numbers to itself.

In the following we use the pair notation for complex numbers and (you guessed it) define two collections of functions that make the Complex Numbers into a monoid.

Let $S'''_{(a, b)}((x, y)) = (a+x, b+y)$,

and $P'''_{(a, b)}((x, y)) = (a*x-b*y, b*x+a*y)$.

The identity is either $S'''_{(0,0)}$ or $P'''_{(1,0)}$

We leave the verification to the reader.

There is an especially simple functor from the Real Numbers to the Complex Numbers, though we use much simplified notation here. Again, the product mapping should be clear. The left downward arrow is actually S''_s and the right arrow is $S'''_{(s,0)}$.

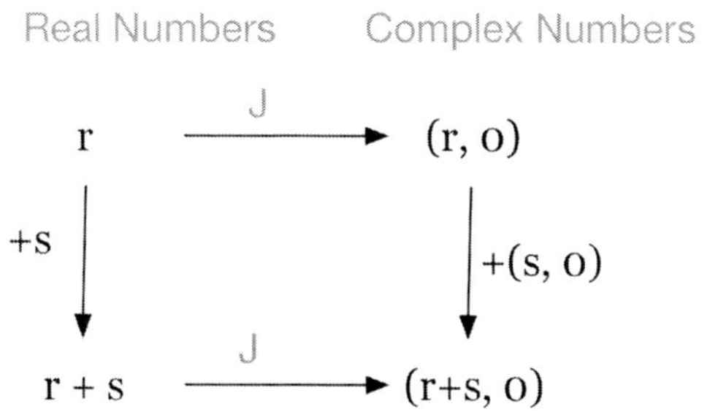

Note that the various functor embeddings we have seen here, such as the functor from the Rational Numbers to the Real Numbers implies that the image of those functors in the destinations are subcategories since they are subcategories in their own right. Moreover, the functors are 1-1 mappings, both of objects and arrows. Thus, since we have the situation that the Rationals are isomorphic to a subcategory of the Reals, we are justified in saying that the Rationals "are" a subcategory of the Reals.

In reality, though, they are more, since other things are preserved as well, such as the commutative and distributive laws (below), which aren't actually part of category theory. This sort of thing is also true in other, much more advanced uses of category theory such as applications to such things as topology.

Polynomials with Complex coefficients can be treated just as we did above for those with Real coefficients.

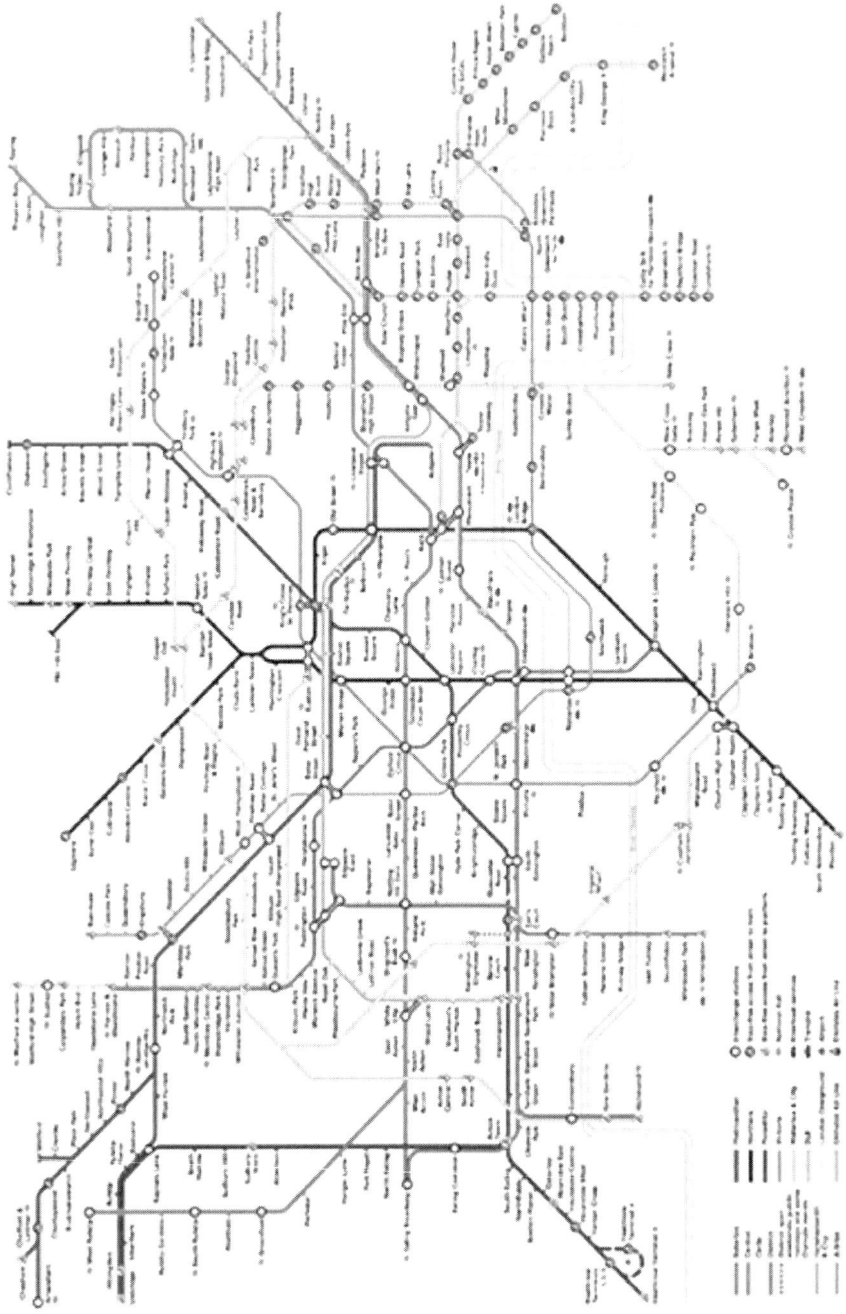

4 Beyond

Looking at the Distributive Law

So far we haven't been able to represent the distributive law in any of these number categories. There is nothing in category theory *per se* that forces or leads to a distributive law, but it can be expressed there through a composition of certain morphisms. It requires a monoid with two operations, which we shall call add and multiply, though their actual meaning is not important. The structure of what we will do is independent of the nature of the operations.

A k-tuple of, say, Integers (or other) is a function from a prefix of the Natural Numbers to the Integers. A k-prefix is just the Natural Numbers from 1 through k: {1, 2, …, k}. Such a tuple or function is said to have *arity* k. So, as a function, a tuple $t = \{a_1, a_2, \ldots, a_k\}$ is just $t(n) = a_n$, for $1 \leq n \leq k$. Since an "object" in a category can be anything, we will let these functions be our objects. Since they are finite, we can represent them by listing their mappings, using subscripts as above. Think of $x = a_n$ as an ordered pair (n, x) and t is a set of these.

We can define transformations (morphisms) of these functions into other such functions. Some of the transformations will change the arity. We will also associate a tuple of arity 1 with its single value for simplicity. We will only need tuples of arity up to 4 for this exercise.

More generally, if $\mathcal{C}$ is any category, then we can define Cartesian products of k copies of $\mathcal{C}$, which also results in a collection of k-tuples. We include k = 1, as well. So, $\mathcal{C}^1$ is just $\mathcal{C}$, $\mathcal{C}^2$ is $\mathcal{C}\times\mathcal{C}$, and so on. Note that forming a Cartesian product doesn't depend on the nature of $\mathcal{C}$. We will show examples from the Integers to give the idea, but it is the operations of the category that actually apply. We now assume that $\mathcal{C}$ is a monoid with two operations: add and multiply.

Let Δ be the transformation that duplicates the last element of a tuple, producing a new one with arity one higher and the final two entries equal So, $\Delta\{1, 2, 3\} = \{1, 2, 3, 3\}$. We could define a more general version Δ_m to be a transformation that duplicates the m'th entry, moving later elements to the right. In that case Δ is Δ_k for a function of arity k. Note that we have, in general, $\Delta: \mathcal{C}^n \rightarrow \mathcal{C}^{n+1}$.

Let $twist_m$ be the transformation that reverses the order of the m'th and (m+1)'th elements. If m is k or greater the let $twist_m$ be a no-op. Then $twist_2\{1, 2, 3, 3\} = \{1, 3, 2, 3\}$. The *type signature*(✦) of twist is: $twist_m: \mathcal{C}^n \rightarrow \mathcal{C}^n$. The operations of $\mathcal{C}$ haven't yet been used.

Let *addL* be the transformation from a k-tuple to a (k-1)-tuple that adds the first two (leftmost) entries, using the add operation of $\mathcal{C}$, producing first entry of the result with the other values reindexed accordingly: $addL\{3, 4, 5\} = \{7, 5\}$. Note that when applied to a 2-tuple there are no more remaining elements, so we just get the sum - a singleton. Also, *addL* is a no-op on a 1-tuple. Function addL has type signature: $addL: \mathcal{C}^n \rightarrow \mathcal{C}^{n-1}$, if $n> 1$. Here *add* is the "add" operation of $\mathcal{C}$.

Let *addR* be the transformation from a k-tuple to a (k-1)-tuple that adds the last two (rightmost) entries producing last entry of the result with the other values unchanged: $addR\{3, 4, 5\} = \{3, 9\}$. Note that when applied to a 2-tuple there are no more remaining elements, so we just get a singleton. Also, *addR* is a no-op on a 1-tuple. $addR: \mathcal{C}^n \rightarrow \mathcal{C}^{n-1}$, if $n> 1$.

The above could be generalized to add_m which adds the m'th and (m+1)'th elements, reducing arity by one and adjusting the entries to the right accordingly. Again, let this be a no-op if m is out of range for the tuple it is applied to. So addL is add_1, and addR is add_{k-1}

Let *multL* be the transformation, similar to addL, that multiplies the two leftmost elements using the multiply operation of the category: Then *multL*{1, 3, 2, 3} = {3, 2, 3}. In general, multL: $\mathcal{C}^n \to \mathcal{C}^{n-1}$, if n> 1.

Let *multR* be the transformation, similar to addR, that multiplies the two rightmost elements: Then *multR*{1, 3, 2, 3} = {1, 3, 6}. multR: $\mathcal{C}^n \to \mathcal{C}^{n-1}$, if n> 1.

The generalization of multL and multR should be clear.

Now, the distributive law in the Integers would say (3+4)*5 = 3*5+4*5. Both of these simplify to 35.

Let's look at this as two series of the above transformations on the tuple {3, 4, 5}.

The first series of transformations will apply Δ, then twist_2, then multL, then multR, then addL.

{3, 4, 5} →Δ→ {3, 4, 5, 5} → twist_2→ {3, 5, 4, 5} →multL→ {15, 4, 5} →multR→ {15, 20} →addL→ 35.

The second transformation will start with the same tuple, but apply addL, then multL. {3, 4, 5} →addL→ {7, 5} →multL→ 35.

Now, using the above transformations as morphisms in the category of Cartesian products of a given category (yes, this needs verification), a monoid with two operations is called distributive, "multiply over add", provided that the diagram commutes.

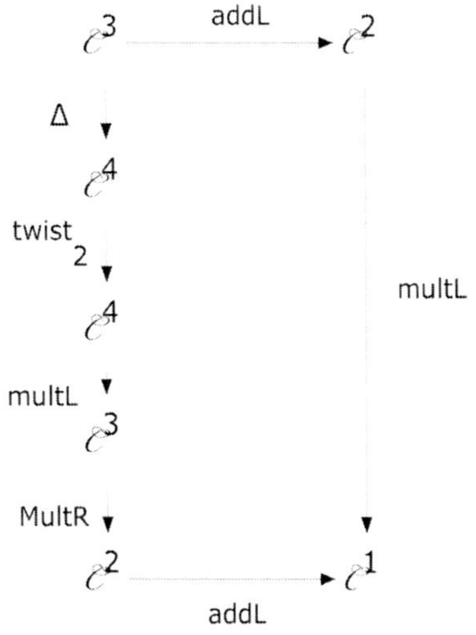

If we take a representative three-tuple of Integers and follow both pathways from the upper left to the lower right we get an idea of how this works. It should seem familiar.

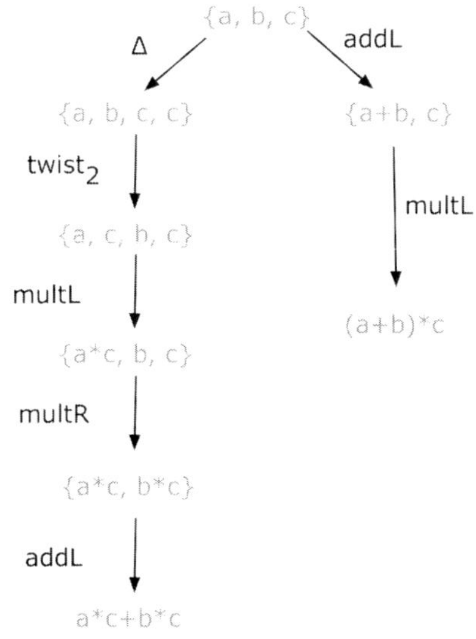

It a computing sense, it is a core idea in *functional languages* that functions can transform other functions into yet other functions as we have done with the tuple functions above. That is actually why the are called "functional". Functions are both operations (like our transformations) and data (like our tuples).

Arrow Categories

Given a category $\mathcal{C}$, we can form another category Arr($\mathcal{C}$), the *arrow category* of $\mathcal{C}$, in the following way. The *objects* of Arr($\mathcal{C}$) are the *morphisms* of $\mathcal{C}$. In general, a morphism in Arr($\mathcal{C}$) can be thought of as an ordered-pair of arrows from $\mathcal{C}$ making a commutative diagram as shown below. Alternatively, a morphism in Arr($\mathcal{C}$) is the actual commutative diagram from $\mathcal{C}$. We will start with the first view and then connect it to the second, standard, view.

Let f:A → B and g:C → D be two morphisms from $\mathcal{C}$. then a morphism from f to g (provided it exists) is an ordered-pair of morphisms (h, k) from $\mathcal{C}$ so that h:A → C and k:B → D forms a commutative square starting at A and ending at D:

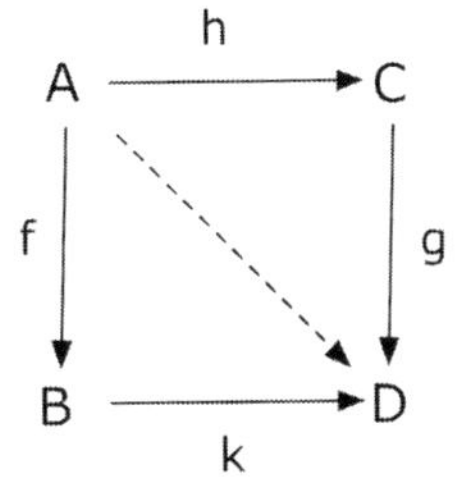

And note that if (h, k) maps f to g, then (f, g) maps h to k as well. So, a diagram like the above defines two morphisms. Well, in fact, this results in isomorphic morphisms as we shall see soon. The pair of identity morphisms on A and C maps h to itself, of course.

Suppose we also have m:C → E and n:D → F. If there is a morphism p:E → F, then (m, n) map g to p and we can compose the above getting a larger commutative diagram. The composition of (h, k) with (m, n) is (mh, nk) as shown below.

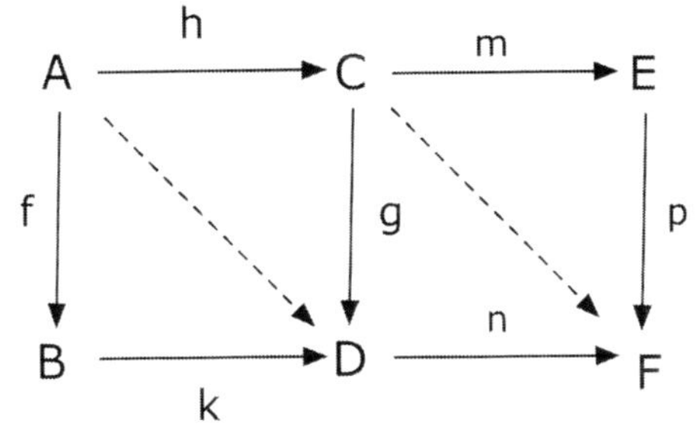

The identity morphisms of Arr($\mathcal{C}$) are mappings of arrows of $\mathcal{C}$ to themselves. The diagram for that, mapping an arrow f to itself, is:

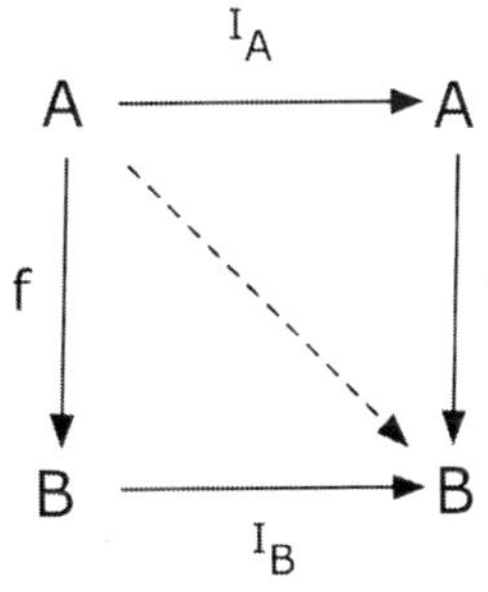

So, it is the pair of identity morphisms of $\mathcal{C}$ that complete the identity morphism of Arr($\mathcal{C}$).

The above interpretation will be useful later, but it isn't the standard one. It even gives an incorrect view of the number of morphisms since some of the ones we have seen are isomorphic. The morphisms of Arr($\mathcal{C}$) are normally thought of as the complete pathways that make up the actual commutative diagrams in $\mathcal{C}$. So, a morphism from f to g in the above would be the pair of paths (no ordering needed), {kf, gh}, the compositions in $\mathcal{C}$ forming a commutative square linking A to D. Likewise a morphism from h to k would be that same pair since both morphisms rely on the same commutative structure.

Using the first formulation, ordered-pairs, if we examine Arr(FACTORS(12)) we find the following. The objects are all of the arrows of FACTORS(12), the union of the HOM sets shown earlier, including all compositions. There are eight non-identity morphisms between these arrows defined by
(a, e)(b) = d, (b, d)(a) = e,
(c, g)(d) = f, (d, f)(c) = g,
(ca, ge)(b) = f, (b, f)(ca) = ge,
(eb, fc)(a) = g, (a, g)(eb) = fc.

We will see in a moment why they have been paired. Note the last two sets of pairs involve compositions in the original category. We can compose (a, e) with (c, g) getting (ca, ge)(b) = f. This composition appears already in our list, of course. Similarly for the last pair.

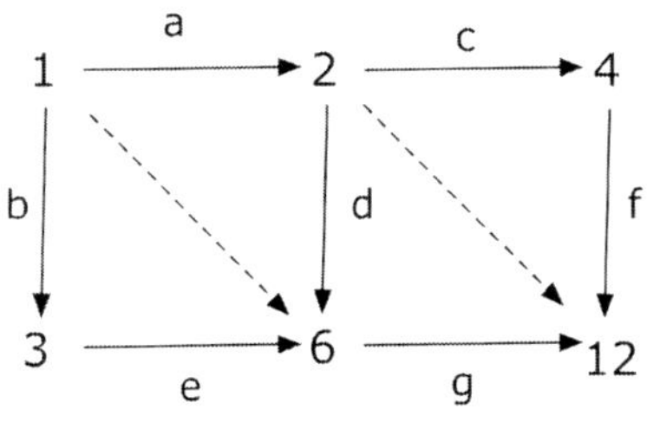

Using the *more traditional formulation (commutative squares)*, first let

the composition arrow eb from 1 to 6 be denoted v,
the composition arrow fc from 2 to 12 be denoted w,
the composition arrow ca from 1 to 4 be denoted x,
the composition arrow ge from 3 to 12 be denoted y.

Then the four, instead of eight, non-identity morphisms, represented as commutative squares, are
{da, eb} which links arrows a to e but *also* links b to d (linking 1 to 6).
{fc, gd} links c to g, but also d to f (linking 2 to 12),
{gv, wa} links v to w but also a to g (linking 1 to 12), and
{fx, yb} links x to y and also b to f (linking 1 to 12).

The fact that the last two both link 1 to 12 is significant as we shall see. But we now count a maximum of four distinct morphisms. There is a distinctive pattern in the above listings. Note that the linking isn't symmetric. There is no morphism from e to a in this category. The arrows of FACTORS(12) point the "wrong" way for that to be the case.

Below are two diagrams for FACTORS(12) redrawn, showing all the commutative connections. Note that {fx, yb} is the composition of {da, eb} and {fc, gd}. Thus, in Arr(FACTORS(12)) we have four morphisms, one of which is the composition of others: four commutative squares. And {gv, wa} isn't like {fx, yb} in that regard.

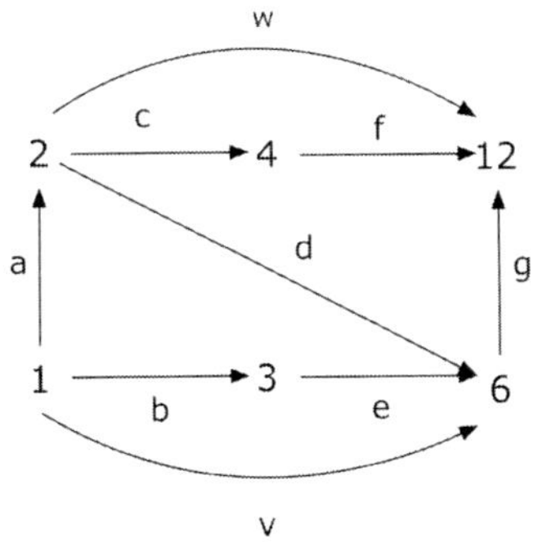

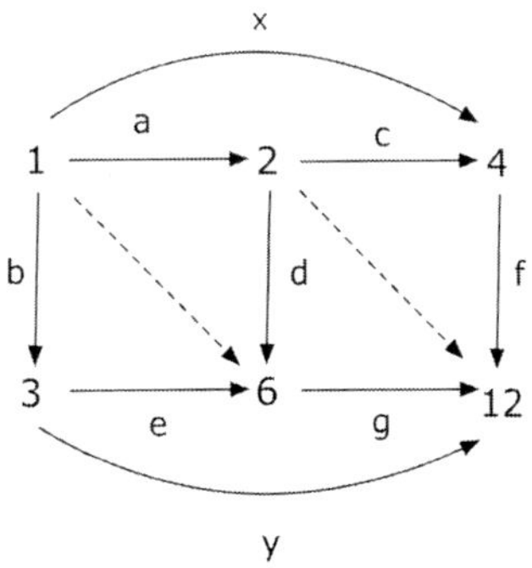

It is important to note that in the two formulations we have different numbers of morphisms, or alternately, in the first view, above, we simply have multiple names for the "same" arrows. So, linking the two ideas the two arrows, (a, e) and (b, d) combine to form {da, eb} of the second formulation. Notice the pattern of the orders of the symbols. For example, where a and e "fit" in the second form. Likewise (c, g) and (d, f) both represent {fc, gd}. Finally (x, y) and (b, f) become {fx, yb}.

Note that FACTORS(24) is an easy adaptation of the above, and does demonstrate transitivity. Give it a try.

We can also now show that morphisms da and eb of FACTORS(12) are isomorphic to each other in the arrow category, hence the "same" in FACTORS(12). To show this, we need two morphisms in the arrow category, one from da to eb and the other from eb to da that compose to the identity (in both directions). Note that both of the following diagrams commute and they also, obviously, compose. The one on the left maps eb to da and the other maps da to eb. And the composition is the identity arrow of eb. The opposite composition maps eb to da and results, also, in the identity arrow of da. We conclude that eb and da are isomorphic, so "the same" in the context of categories.

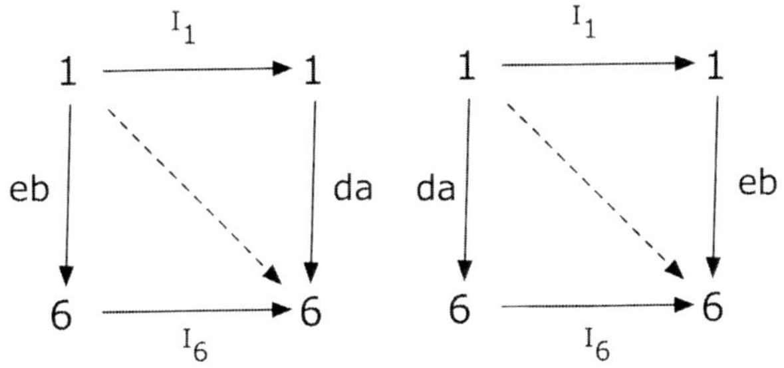

The same will be true for the other sets of pairs: fc is isomorphic to gd (linking 2 to 12), fx is isomorphic to yb (linking 1 to 12), and gv is isomorphic to wa (linking 1 to 12). This also resolves the difference in the arrow "counts" between the two different views given above, and also lets us look again at the Hom sets of FACTORS(12). You might want to convince yourself that the three pathways from 1 to 12 are all isomorphic. Hint: consider wa and yb as well as gda. That last pathway hasn't yet been seen here. You should conclude that there are three, not four, distinct morphisms. And think about the implication of squares like the above, and not just for a FACTORS category, but in general.

Arrow Category For the Whole Numbers

Earlier we looked at the Whole Numbers under Addition as a category in two different ways. In the monoid view, the morphisms were the Whole Number values and the compositions were addition. So, the 3 arrow, composed with the 2 arrow gives the 5 arrow. We call this the monoid view.

In the other view, the Whole Numbers were the objects of an infinite (in objects) category, and the morphisms were generated by functions, such as $f_2(x) = x+2$ and $f_3(x) = x + 3$. Composition was function composition and f_2 composed with f_3 gives f_5. We call this the functional view.

Here we will show that the functional view is actually the arrow category of the monoid view. Additionally, it gives a bit of insight about how we get from the Whole Numbers to the Integers. It will be helpful to use the first, ordered-pair, formulation of the morphisms of the arrow category.

First, let's look directly at the arrow category of the monoid. Suppose we look at the 3 arrow and see if and how it might have a morphism to the 5 arrow, keeping in mind that composition in the Whole Number monoid is just addition.

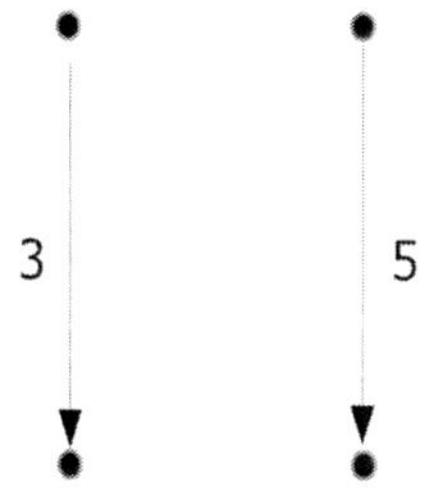

We can complete this diagram with the 0 morphism across the top and the 2 arrow across the bottom. It will then commute. 0 + 5 = 3 + 2. This morphism is the ordered pair (0, 2), of course. But the 1 and 3 arrows work just as well; 1 + 5 = 3 + 3. And the (2, 4) morphism and the (3, 5) morphism and, and, and. *But this set of ordered pairs is exactly f_2* from the functional view.

The same will be true if we replace the 3 and/or the 5, say with 2 and 9. The morphism between these arrows will the ordered-pairs that define f_7.

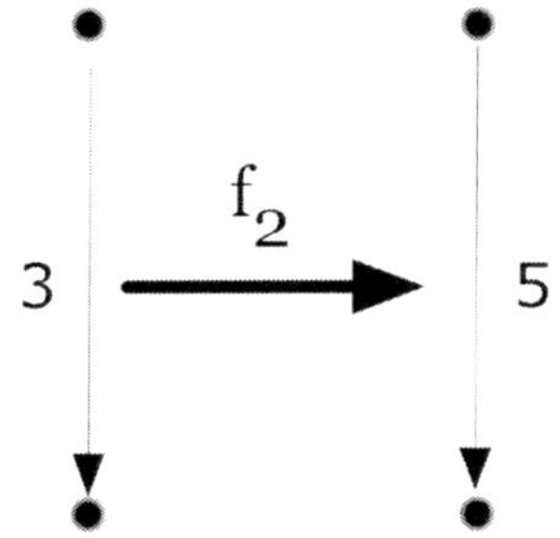

Or, more traditionally,

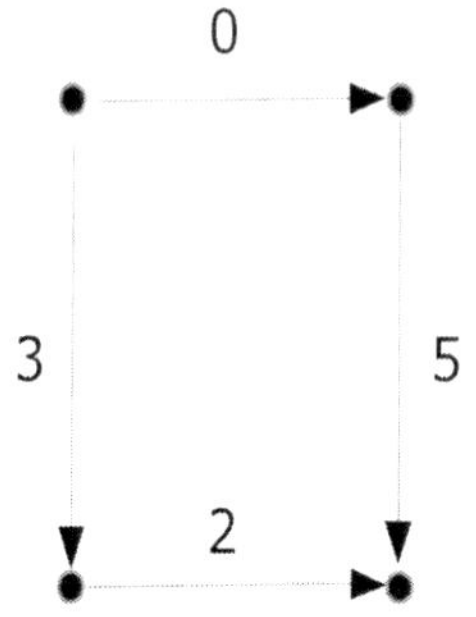

But there is another interesting relationship here. Suppose we want to reverse those two arrows and try to map the 5 arrow into the 3 arrow. If we put the 2 arrow at the top and the zero arrow at the bottom, morphism (2, 0), we get a commutative diagram, hence a proper morphism. To visualize this, you can actually just draw arrows pointing to the left instead of the right as we usually draw them here, of course. But (3, 1) works just as well as does (4, 2), and on and on. But this is just f_2^{-1}, the inverse of f_2. Now recall how we moved from the Whole Numbers to the Integers with an equivalence class of ordered pairs, just like those of f_n and f_n^{-1}.

A Last Look at the Whole Numbers as a Category

Define a category W with two objects, b and x. There are four sorts of arrows: the identities on the two objects, a morphism, s, from x to x that is *not* the identity on x, along with a morphism named 0 from b to x. The identity morphism on x will be denoted I. Object b is where we *begin*.

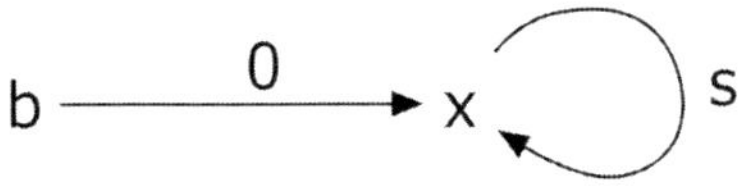

Such a figure represents something similar to a NNO, except that a NNO exists only within a set and here x needn't be a set, nor b a terminal object. Our interest at the moment is to look at induction in this category and, in particular, how the sum operator was defined in the Whole Numbers in The Book: a + 0 = a, a + b′ = (a + b)′. Recall that every Whole Number x has a successor x′, so if a + b is defined, then its successor (a+b)′ is used to define a + b′, by induction on b.

Again, we can, starting with b, represent numbers by composing s with itself such as in the following diagram. Three compositions of s can represent 3, provided we start at b. Thus, s is like the successor operation in the Whole Numbers. Technically the long arc below is $s^3 0$, the composition of four arrows.

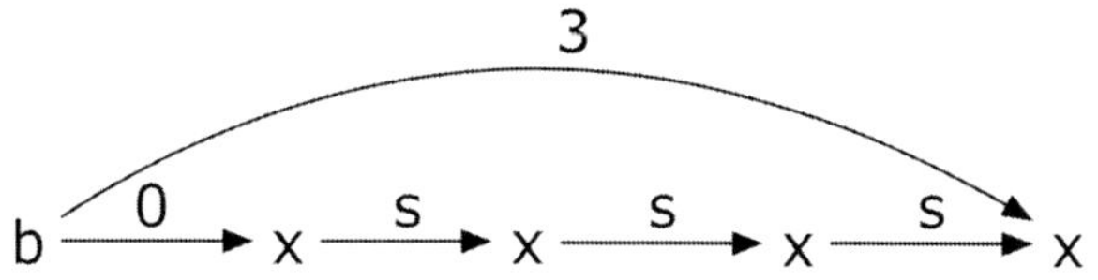

However, we can also think of three as represented by three compositions of s following an application of the identity I of object x, as in the following:

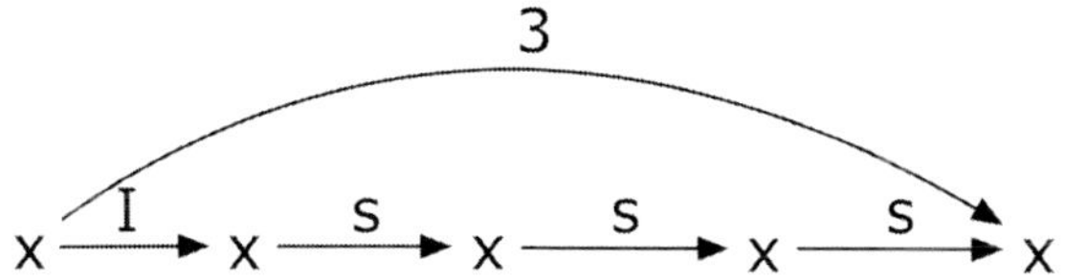

And, since I doesn't "change" anything, we can compose the above two diagrams to get six. Since the identity can serve as a separator, we represent three plus three equals six.

Also, the successor operation, "prime", of the Whole Numbers, is represented here by "one more" application of the s arrow. So, below is three again (the upper arrow, represented at the successor of 2 (the lower path).

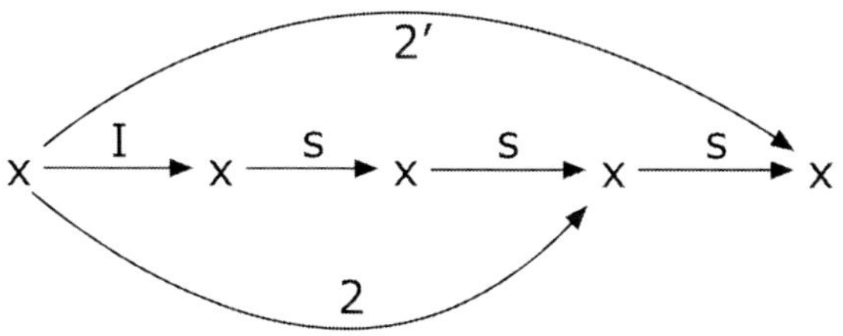

Next, let's look at an example of using composition of these diagrams to illustrate the definition of the sum. In particular we look at 3 + 2′ and how it relates to 3 + 2 and to 3 + 3.

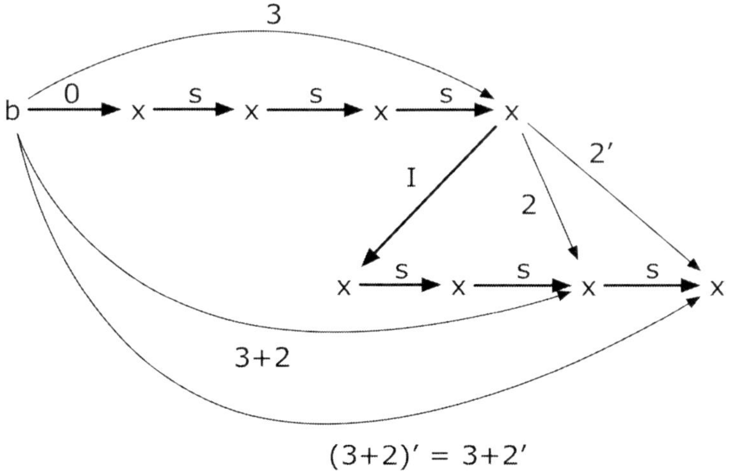

The upper arc shows the representation of 3. Then, applying I takes us to the lower diagram. The inner arc on the lower part shows the sum of 3 and 2, since the diagram composes the three arrow at the top, with the two arrow from the x at the end of the top of the figure, starting with I from there. And the final application of s implies 3 + 2′ since it is one more application of s to the already applied 2 applications of s along the bottom; Follow the upper arc and then the 2′ arrow to reach the end. But that same application of s transforms the 3+2 arrow (arc) by one more application of s, representing (3+2)′.

Finally, there is nothing magic about 3 and 2, so, simply by relabeling we get the final diagram.

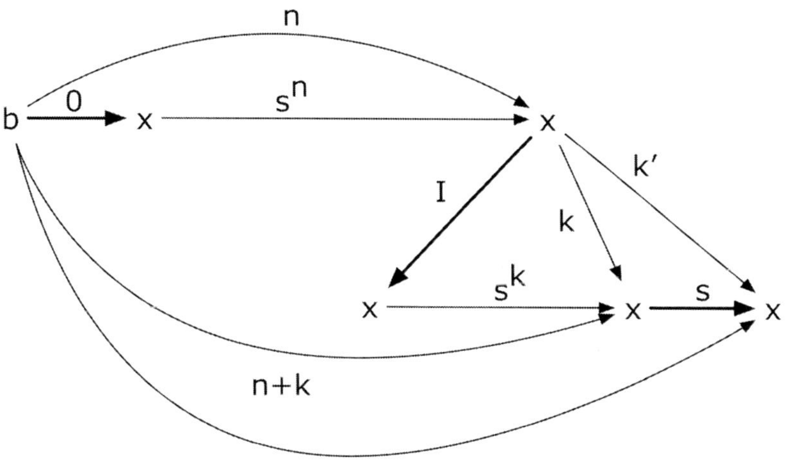

(n+k)′ = n+k′

We let n applications of s represent n, either from the beginning, b, or from an application of the identity, I. It is instructive to show how this figure collapses if either n or k is zero.

It should be clear how to create an invertible functor between W and any NNO.

Which, if any, of the following four categories (as graphs) best represents the Integers? Why? Say why they work or don't work. Are there conditions under which one or more of them does work?

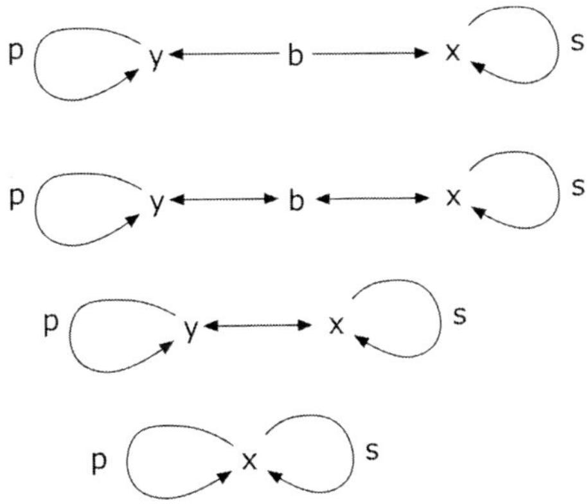

And, note that this sort of thing fails for the Rationals, Reals, and Complex Numbers due to the denseness of those structures. There isn't, for example, a "successor" or "predecessor" in the Rational Numbers.

Z80 Architecture

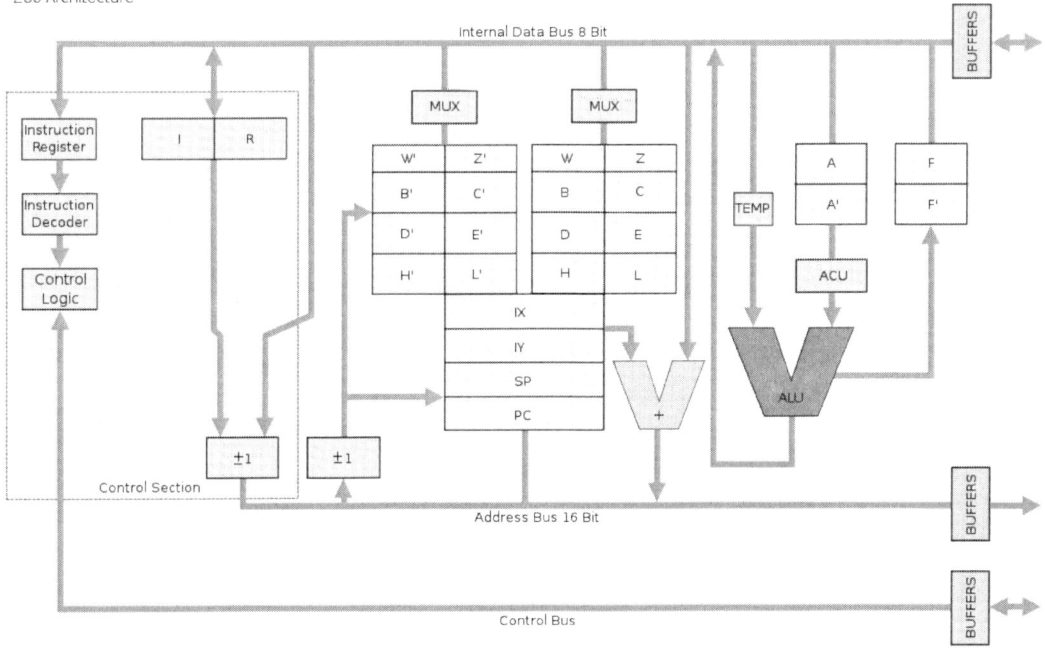

5 Categories and Computing

Types and Hierarchies

Categories appear in computing in several ways. They generally involve data types, both those built-in and user defined types. Different languages define different kinds of built-in types. Some simple like *Integer*, but also things like tuples and lists.

In Object-Oriented(+) programming languages, users can create new data types as classes. Classes, using inheritance(+) (subclassing), have some sort of hierarchical structure. In a single inheritance language like Java(+), with a required top level type, *Object*, concrete classes, hence types, seem to have a strict tree-like structure. Thinking of all of the relationships as arrows pointing upward toward the root, they form a graph, hence a rather boring category. The classes are the objects of the category and the subclass relationship generates arrows. Instances of the classes are also called objects, but that is a rather different meaning.

In multiple inheritance languages like Python(+) the class-type structure is still hierarchical, but can consist of several independent or intersecting hierarchies.

But it is more complicated than that. In Python a class can have several supertypes, so the hierarchies can intersect forming more complex graphs. But still, all of the arrows point in one direction. This type structure is true even in Java, however, since *interfaces* define types and a class can have (implement) several independent interfaces. So, an inheritance graph in a simple program might look like the diagram. This could be true even in Java, if A, C, and

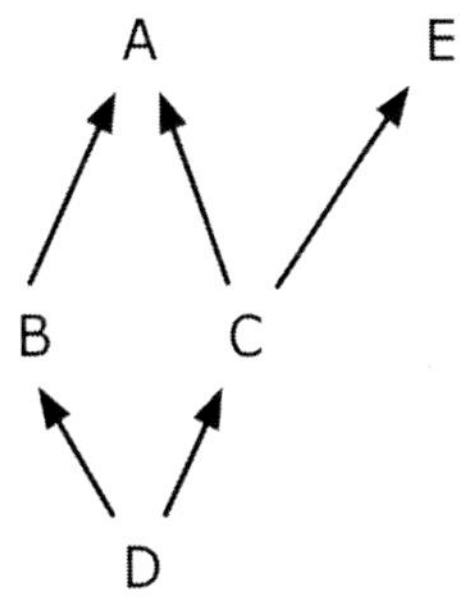

E are interfaces, but is more typical of a multiple inheritance language.

Functions

The second, very different, way in which types and categories work together is with simple functions. Here it is the types that are the objects of the category and functions with an argument of one type that returns a value, possibly of another type, represents an arrow from one type to the other. For this purpose, there might also be a "dummy" type to correspond to "void" or "none" for those functions that don't return a value. They might be arrows from "int" to "void", for example.

The length function might have *String* as source and *Integer* as destination in some language. Thus, a simple arrow from String to Integer, though there may be many other such arrows as well.

Functions that have multiple parameters[+] can be thought of as having a tuple for an input parameter, even when the syntax doesn't suggest that.

Note that functions can be composed associatively and the identity function between types is always available, if seldom used. f(x) = x is independent of type, as long as the language lets a given type be a returned value (there are exceptions in some older languages).

The above categories are usually defined statically, in the text of the program. But there can also be dynamic aspects that we will see in a bit.

There are, however, two basic kinds of functions. Some don't have side effects and, like a mathematical function, return the same information every time they are called with the same parameter(s). There are also functions with side effects, perhaps nothing but side effects, such as a print function. These ideas can be mixed in a given function, but it is generally thought unwise to do so. Functions without side effects (*pure* functions) are *accessors*[+] that produce information, but don't change it. Functions with side effects are *mutators*[+] that change the state of a computation but don't return values. If a program has a lot of functions that do both, it is likely to be difficult to understand and modify.

Paradigms

There are differing programming paradigms. A paradigm(+) is a way of thinking. A language will generally try to implement a single paradigm (C++(+) is an exception). Haskell(+) implements the functional paradigm. Python is (largely) object-oriented. Ruby(+) is purely object-oriented. The problem with paradigms is that they require a way of thinking about the nature of computation and, as such, it can be difficult to switch languages, and, especially, to learn a language that uses a different paradigm than the one(s) you are familiar with.

In particular, the functional paradigm advantages pure functions; those without side effects. The earlier section on polynomials discussed functions that produce other functions. Eventually, though, a program needs to produce output - some effect of the computation. Haskell has techniques for controlling such side effects and bringing them into a purer functional form.

On the other hand, object-oriented languages prefer mutators, rather than accessors for the methods defined in a class. When you invoke a method of a class (send a message to an object of that class), you are encouraged to "tell, don't ask"(+). In other words, messages instruct the object to carry out some task, rather than produce information that is used to carry out the task elsewhere.

Thus, functional programming favors non-mutability, while object-oriented programming seems to favor mutability. But even that is subtle. It is possible to create objects in object-oriented languages that produce services without themselves being modified. These immutable objects are favored, since a program with a high percentage of immutable values is easier to reason about.
But in both paradigms, a function or a method can produce a cascade of

calls/invocations. In functional programming, a (pure) function may well call other functions to help it build up the return value. In object-oriented programming, a message to an object to "do something" will send messages to other objects to carry out parts of the task. But the *ask vs tell* can be disconcerting to a new programmer in a new language.

And note, that the call/message actions of a running program are actually arrows between types, and the execution will produce a different path each time it is run with new data inputs. But a message sent from one object of type A to another of type B is an arrow from A to B. This is normally quite different from the static nature of the type hierarchy.

For example. In the Monty Karel version of the Karel programming systems, intended for beginners, the basic sort of robot, UrRobot, has no accessors at all. You can "tell" a robot to *move* or to *pickBeeper*. It will, behind the scenes, interact with a graphics system to produce changes in the screen representation. The student programmer creates subclasses of UrRobot to extend the capabilities defined by the basic five actions. The figure below shows a single robot in a world with a few walls and four beepers. The task is to pick up all of the beepers. The program might not depend on the specific configuration of walls, with some being longer than the example.

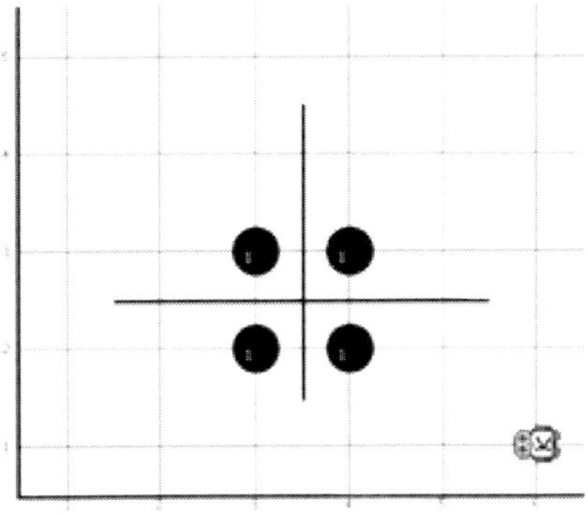

Later in the book, another class, Robot, is defined, extending UrRobot and adding some "sensing" capabilities. These are all accessors, returning Boolean (true-false) values about the state of the robot: *frontIsClear*, *nextToARobot*, etc. These methods have no inputs (other than *self*) and the result depends on the current situation of the robot within its world. The student, again, extends this class with additional capabilities, mostly mutators. There is no way, in this model to ask a robot which *street* it is on, nor which direction it faces. It can, however, ask if its *frontIsClear* or if it is *facingNorth*. But the boolean methods are sufficient to turn the system into a Turing Machine, permitting it to carry out arbitrary computation tasks without explicitly using integer variables. The only "primitive" values in the programming model are booleans and robots of various kinds (well, beepers, too). But the robots can interact with the world, by sensing *walls*, *beepers*, etc. The model supports programming with recursion, but such things as "counting" by changing the value of an integer variable is not used.

Metaphorical Categories

Metaphor and analogy work well in designing and in using programming languages and systems. If a user has an idea about what is "really going on" that is consistent with what happens, though not necessarily accurate, they can make predictions about the effect of changes. Object-oriented languages have a basic metaphor that objects (things with a class type) are real "things" that react to *messages* sent them.

Even more important, a programming system, built on top of a language can have additional metaphors that aid understanding. The Karel system, for example, induces the student to think of the robots as real "things", even though real robots in a real world don't act quite the way the robots in this system do. For example, *turnLeft, exactly 90°,* is more or less impossible to implement accurately and consistently in a real world robot. Approximations and correction is the name of the game in the real world, but the metaphor works in the Karel system.

This brings us to a different sort of category in programming. If one object "knows about" another, say by having a reference to it. Then the first can send messages to the second. Here the objects of the category *are* the objects of the system, and those messages are morphisms - connections. The object receiving a message can pass other messages to other objects (or to itself), extending the connection (morphism) graph of a running program, or a static program if you consider only the types and not the objects. Thus, a running program will construct a category as it runs, and the metaphor of the system lets the programmer understand the general structure of this category implicitly.

The metaphor in the Karel system seems to be that the robots "understand" their world and can interact with it. In reality, the software itself encourages that view, but doesn't implement it. In fact, it is the World object that knows about the robots, not the other way round. But that detail gets in the way of the user programming a robot to pick up a line of beepers. Metaphorically, however, the user program constructs a system of connections (morphisms) between objects (beepers, walls, robots, directions, etc.): a category.

The Karel software system is actually built using a collection of Software Design Patterns[+]. One of the most fruitful of these is the Observer Pattern[+], that lets an observer (the world) know whenever an observed (a robot) executes a method. This establishes arrows between the world and the robots that are exploited as the program executes, updating the graphics system. In fact, the graphics system itself, observes the world, which is the underlying model of the system. This idea of an observer is another level of metaphor, of course.

The major metaphor in functional programming is more like a manufacturing facility that takes inputs and produces outputs of arbitrary complexity. Along the way, it might use subcontractors to supply "parts" to be used in the final output. A modern functional language can also transform what would be mutators in other languages into (seeming) accessors of complex data. For example, a function might produce both some desired output (a widget) along with a description (a string) of it that might be used in "advertising". If the Karel system were built this way, the output of a command to a robot might, in principle, be that same robot along with a new "picture" of the world as affected by the command. The programmer then "thinks" in terms of transforming inputs into outputs. The pictures could be collected and made into a movie.

If one is building a software system for use by others it helps to develop a user metaphor with certain implied connections between components (arrows between objects) that are natural to the user. But the architecture of the system can be very different from the metaphor for its use. For example, the Karel system is built using Model-View-Controller[+] architecture with the world as the model, the visible screen, as seen above, as the view and the student programmer's program as the controller. This implies quite different connections between objects, so a different category, than that in the student's understanding. The architecture needs to support the metaphor, of course.

An Example from Java

There is a simple, but interesting, application described in *Beyond Karel J Robot* that shows the difference between the user metaphor and the architecture, both of which can be represented as categories.

The *French Military Game*, also called by various other names including *Hare and Hounds,* involves an attempted escape and pursuit. In the version discussed, a spy called The Fox is trying to evade a team of three Police. The game is played on a board like the following:

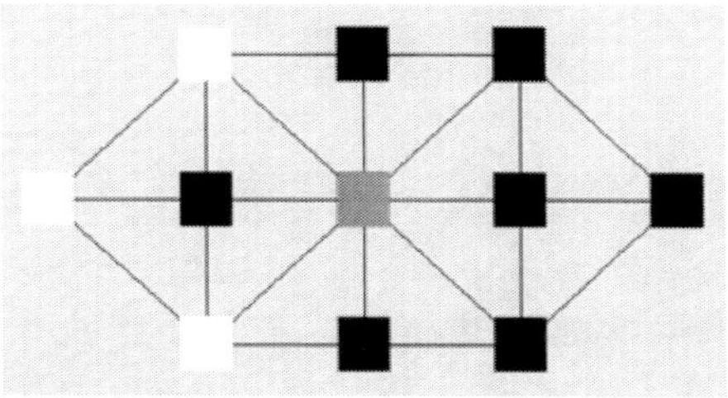

Initially The Police occupy the three leftmost positions as shown in white. The Fox is at the center, in grey. The human player controls The Police, who move first and can move one marker per play, but only to the right or vertically along the marked lines. This is, of course a directed graph between the corners which can serve as objects in the category they use.

The directed graph defining possible Police moves is shown below:

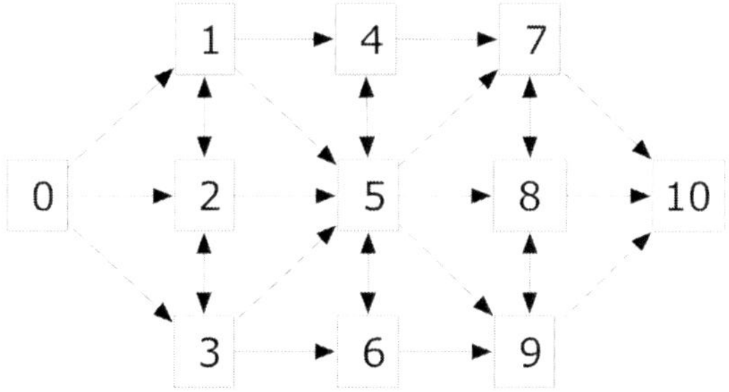

The Police positions are initially locations 0, 1, and 3, with The Fox at 5. The Fox can ignore the direction of the arrows, or, alternatively, its graph has all double arrows between the linked positions.

The Fox can move in any direction, one place per play, back and forth with The Police, but not to a location currently occupied by The Police. The objective of The Police is to trap (not capture) The Fox leaving it no move. For example, if The Fox is at 4 with The Police at 1, 5, and 7, then it is trapped and The Police win. The Fox can also be trapped at 6 and at 10.

The objective of The Fox is to move "behind" The Police to the left, eventually occupying the leftmost square, evading The Police. The Fox will also win if it evades The Police for 20 moves.

Thus, the metaphor of the game, seen by the human player is a board game with movable pieces and a set of rules.

In the program described in the book the computer plays The Fox, though that part could also be played by a human. Game Theory(+) says that The Police can force a win by playing wisely. Initially, The Fox will play naively, but it remembers all previous board configurations and whether they led to a win or a loss. Thus, it "learns" to play better. The actual amount of information required to do this is quite limited. Eventually it becomes very difficult to out fox The Fox. Moreover, it is possible for the game to remember all past plays so that restarting after quitting starts out a new game with a smarter (foxier) Fox. The architecture of classes for this program is as follows with the arrows implying that the source requires information from the destination. There are no subclass relationships here.

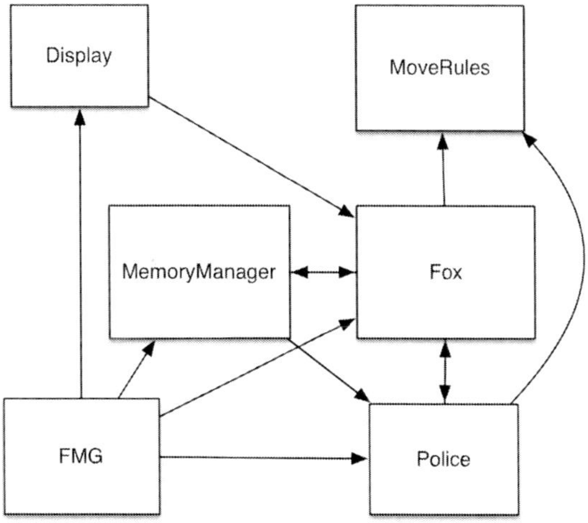

Information moves within the game during play in the direction opposite the arrows. The MemoryManager, for example, queries The Police for information. The initial object in this category is the game itself: FMG. MoveRules is terminal. There are several other data types and classes required (and some implementation options), especially for the MemoryManager which keeps track of all configurations: where every token is after every move over a series of games. This is the game *history*.

There is an interesting connection to The Book as well. You can encode the position of the three Police tokens in a single eleven bit number with exactly three nonzero binary digits (bits). There are exactly 165 of these values: the number of ways to choose 3 from a set of 11, as taught in Finite Math(+). You can encode The Fox position as a number 0-10. Thus, a given configuration is a pair of numbers. The initial position of The Police is $1*2^0 + 1*2^1 + 1*2^3 = 11$ (base 10).

Thus a two dimensional array/matrix of size 165 by 11 will hold any given game configuration, with rows representing Police positions and columns being The Fox. With all matrix cells initially 0, increasing the value at a cell if that position leads to a win and decreasing it for a loss gives The Fox the information it needs to play future games based on past ones. This matrix is updated at the end of each game and used by The Fox to guide its play.

Each single game's history is a 20 by 2 matrix with a row for each pair of moves (maximum 20) and a column for each player. The Police column has one of those 165 values and The Fox column a number between 0 and 10 representing its position at the end of the move. This matrix is scanned at the end of the game to update the history matrix.

The final piece that puts this together is the mapping between The Police positions and the rows of the history matrix. The numeric values, described above, for the three Police run from 7 = 111(binary) to 1792 = 11100000000(binary), but they are only 165 of the 1785 possible values. To manage this an *indexing* array contains the possible Police positions in some arbitrary order. The order is immaterial, as long as it is used consistently. This indexing array is computed once when the game starts. The *Beyond Karel* book uses a naïve ordering as the book is designed for beginners and so sorting(+) and binary search(+) aren't discussed. It is possible (and just as efficient) to generate them in increasing numeric order, however, and make searching for the relationship a bit faster during play. The search looks through the array for a value and returns the index at which it is found. This is, then, the row number of the history matrix. Binary search for a value in a sorted array with 165 values requires a *maximum* of 8 probes, while linear search in the array requires an *average* of 82.5. That seems significant, but, in a game with one human player, the difference won't be noticed.

A HashMap(+) is another good way to store the history and has definite performance advantages over the matrix. In particular the computation of The Police position can be used directly even though they don't form a contiguous range. This avoids the complication of the indexing array. This is beyond the scope of the referenced text, however. But the exercise of building this game is fruitful in exploring Data Structures(+).

There is another way to look at a program; the flow of control as it executes. This technique, flow charting, was once a main design tool for programs. One of my mentors, David Gries, often described them as Flaw Charts. The problem in using them for design is that they have too many degrees of freedom, since you can draw arrows indiscriminately, which can lead to incomprehensible programs. Structured Programming(+) was an early response and it led to better languages and Design Patterns(+). But a flow chart is a graph and hence generates a free category.

Here is the rough outline of the flow of control in FMG, though it omits some steps.

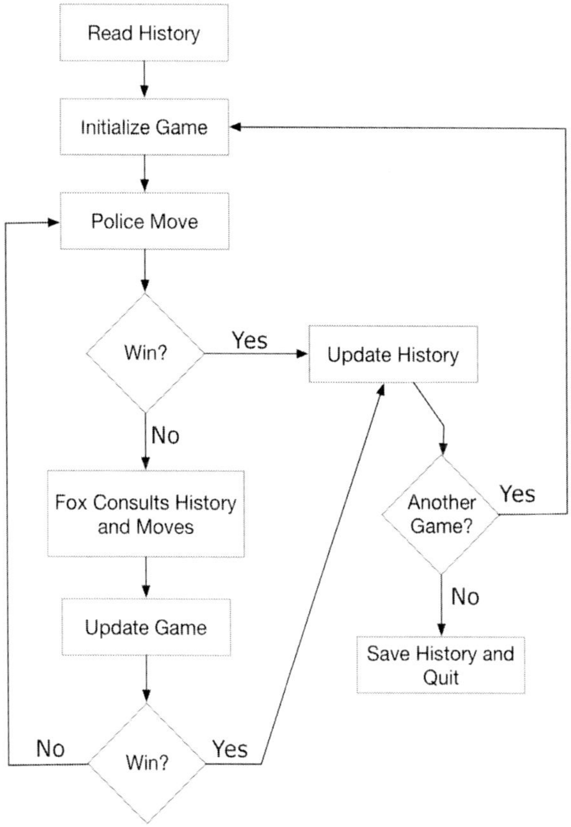

The human player might not include all of the above in their (metaphorical) view of the game. Their view is likely to be much simpler. Think for a moment about what they believe the flow chart to be, observing the game as a player, but from the outside. How is their view related to the above?

Contrast the information shown here with that of FACTORS(12) and the class diagram for FMG shown previously. In FACTORS there is no concept of flow, just a simple relationship. The class diagram shows information flow and their direction, but not the sequencing or timing of the flows. This flow (flaw) chart, however, is all about the timing of things but leaves question of the source of the information unsaid. Here, an in-arrow implies previous action and an out-arrow to a cell (object) implies later action. While, being graphs, all three generate free categories, a very different type of information is carried by each. This illustrates, though still incompletely, the power of category theory to represent information and brings together common elements of very different mathematical and computing concepts.

The reader might also explore building something like the tic-tac-toe game along the same lines as here, with computer playing one side and a human the other. The same flow chart might work. How complex would a game need to be so that a different flow is needed.

Skeleton Subcategories

In the French Military Game, the two players have different sets of connections of the board locations along which they can move: different categories. The arrows shown on the diagram of the board, above, show The Police moves. The Fox can treat all of the arrows as if they were double ended (actually, two arrows, one in each direction).

For The Police, some of the board positions, the objects of the category, are isomorphic. For example, 1, 2, and 3 are all isomorphic with the two connections between 1 and 2, for example, being inverses.

In a category with isomorphic objects there is the concept of the skeleton of the category. A skeleton is a subcategory that has no isomorphic objects but otherwise contains *all essential* elements. In other words, you can select one representative of each collection of isomorphic objects along with the necessary morphisms to make a subcategory. For The Police category 0, 2, 5, 8, and 10, along with the direct arrows between them is a skeleton. But so is 0, 2, 5, 7, and 10, with appropriate arrows. So, skeletons aren't unique. However, they are isomorphic as categories.

But, while you can delete one of the arrows between objects such as 4 and 5, etc. making them not isomorphic, you get a subcategory, but not a skeleton.

A skeleton is, in some sense, a minimal subcategory that keeps "essential" elements. Such a thing collapses isomorphic objects into a single representative. If you were travelling along the graph, from 0 to 12, there are several paths you could take that don't really differ. 0, 1, 5, 9, 10 is as good as 0, 2, 5, 8, 10.

The following is one skeleton of The Police category:

In particular, you can get from 0 to 1 in the original in just the "same" way you get from 0 to 2 here. 1, 2 and 3 have been collapsed into a single object, carrying their arrows with them.

But, keeping "essential elements" requires that a category and its skeleton be isomorphic. The object groupings we've seen are easy to show are isomorphic, but that isn't quite enough. The arrows also need to be isomorphic if they are to be "collapsed". The arrows from 0 to 1 and from 0 to 2 are also isomorphic, as are the other "collapsed" arrows. To see this for the two mentioned arrows, examine the two diagrams below where we have labeled the arrows:

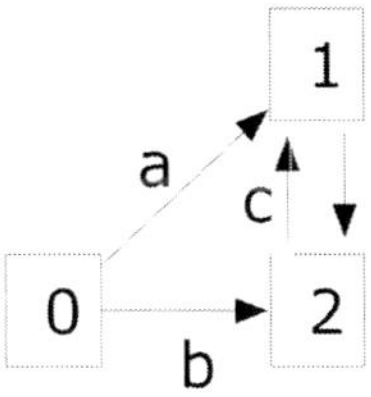

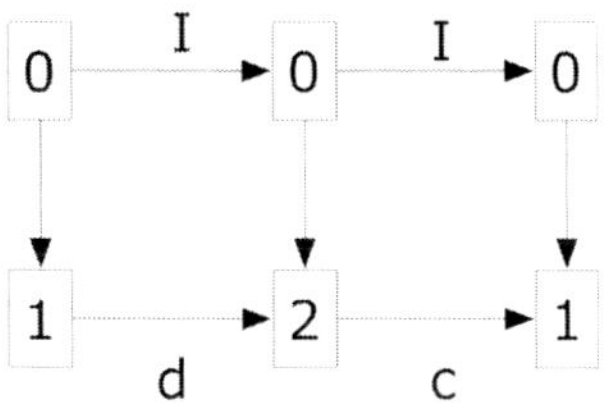

The other arrows such as the five from any of 1, 2, and 3 to 4, 5, and 6 can similarly be shown to be isomorphic. Thus, we have a skeleton.

The following is not a skeleton, even though it has no isomorphic objects and is a subcategory:

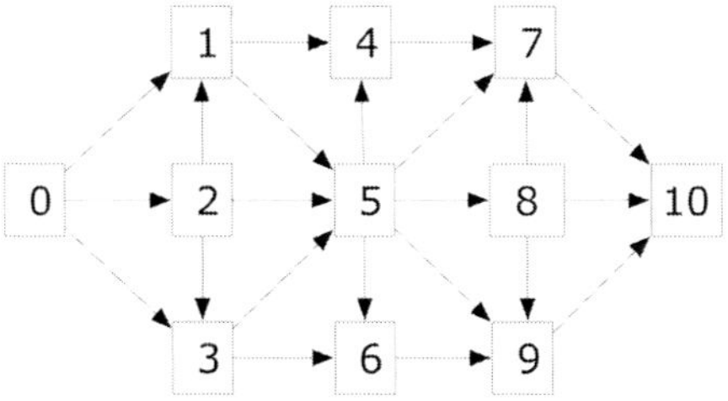

While skeletons can be a useful concept in category theory, it isn't especially useful for The Police to build a strategy around.

What is the skeleton of the category of Fox moves?

In SET(S) and FINITE(S), sets of the same cardinality are isomorphic. So, a skeleton for either is a collection of sets, one of each cardinality. For FINITE(S), this subcategory is isomorphic to the Whole Numbers. It is a bit more complex if infinite sets are included, but the same idea applies. But, even in the finite case, this requires showing that, as morphisms, all functions from a set A to itself are isomorphic. It is a simple exercise and worth drawing the required diagrams. The result depends only on the composability of functions and the fact that all functions here are total.

In the numeric categories that form the core of this exposition, there are no isomorphic objects. So, for example, the Integers form their own skeleton.

Just for fun determine which objects in the category defined by the following graph are isomorphic. Determine which arrows are isomorphic. Determine a skeleton for this category.

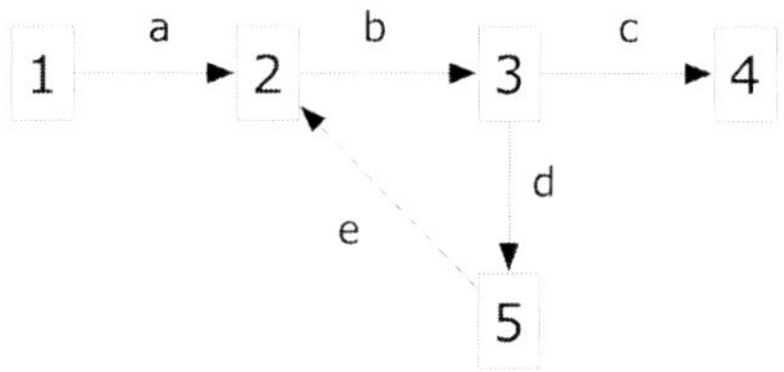

If you consider the arrows here to be an ordering, then this structure is an example of a preorder(+), but not a partial order. It is reflexive and transitive, but not antisymmetric.

Take the potential 165 numeric values of Police positions in the French Military Game as objects. Let there be a morphism between two of these objects if The Police have a move from one of these to another. Thus, there are morphisms from 11 (the initial position) to 7 and back again. There is a morphism from 11 to 14 but not the reverse. Write a computer program to list all of the morphisms in this category. What is a skeleton of this category? Which objects and which arrows are isomorphic?

Ponderables

Can you rewrite Peano's axioms in the language of categories: objects and arrows?

Can you formulate (or not) any of the following as categories, perhaps monoids? If not, decide the various ways they fail.

1. Square matrices of, say, Integers, of a given size with addition and matrix multiplication as operations.

2. Arbitrary matrices with addition and multiplication.

3. Differentiable Real valued functions of a Real variable with the first derivative as the morphism linking two functions.

4. Three dimensional Real vectors with the cross product as an operation.

5. Infinitely differential Real valued functions of a Real variable with any n'th derivative relationship as the morphisms. Include 0'th derivatives interpreted as the identity function.

6. Last, but not least: Rubik's Cube:

One More Thing

While you are pondering, think about the following. Yes, it may seem a bit weird, but it also tends to unify a few ideas.

As usual { } stands for the empty set and 0 stands for the "base" value of the Whole Numbers, with 1 it's successor.

But suppose we make the following alternate definitions of the Whole Numbers

0 = { }
1 = {0}
and in general:
n = {0, 1, … n-1}

Think about what that means, especially in terms of the Peano Axioms as discussed in The Book, but starting with 0, instead.

Also, think about it in terms of the relationship between a finite set and its cardinality.

Note that we "define" a number as a set of sets. What is the set representation of 3? How does that compare with that of 4?

What is the minimal number of characters needed to represent n in the set representation, not counting spaces?

Can you write a program (recursive or iterative) to write out the set representation of n?

What interesting things can you come up with?

Foreshadowing

Let's look ahead a bit, suggesting further study.

If you take a course in, for example, Abstract Algebra or Group Theory, categories might come up as a topic or not. But if you take Algebraic Topology or Algebraic Geometry, then it almost certainly will, since composeable and commutative diagrams are used there to express important concepts.

But even in Category Theory itself, there is much more. In particular, we have restricted the discussion here to *small* categories in which Set Theory is sufficient to express the concepts. Even an infinite category like SET(S), for an infinite set S, is small since Set Theory is sufficient. But if you remove the restriction that we start with only subsets of a given set, you have a category SET that is a collection (but not a set) of *all* sets with all functions among them as morphisms. But even this category is *locally small*, since for any two objects in SET (i.e. sets) there are only a *set's worth* of functions among them.

But you can go further, as we've only hinted above. Suppose you take categories, themselves, as objects and functors as morphisms. This is category CAT. Now you are truly large. There is a lot to learn.

Go, do it.

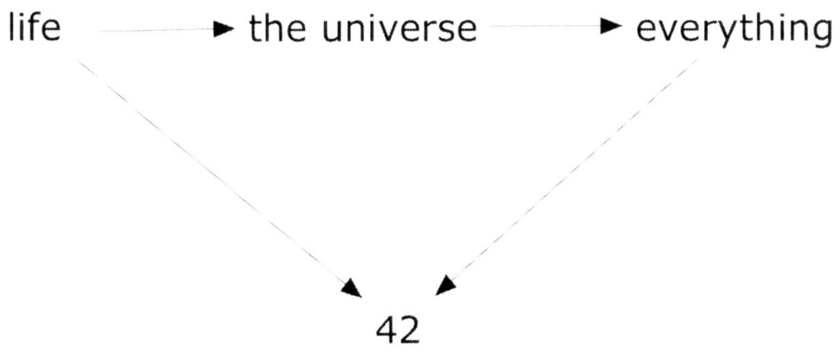

Happy computing!

Symbols Used

$\in$ element of a set: a $\in$ A

$\notin$ not an element

$\subseteq$ subset of another set

$\supseteq$ superset of another set

{ }=$\varnothing$ the empty set

$\exists$ there exists

$\forall$ for every

$\cup$ union of two sets

$\cap$ intersection of two sets

| used variously for "divides" or "such that"

$\cong$ is isomorphic to; a @ b

g$\circ$f composition of f then g

$\times$ cartesian product (or more generally, products)

Bibliography

Awodey; *Category Theory*, Second Edition, Oxford University Press, 2010.

Bergin; *Number and the Meaning of Mathematics*, Slant Flying Press, 2023.

Bergin, Stehlik, Roberts, Pattis; *Monty Karel*, 2008, ISBN 0-9705795-2-7.

Bergin; *Beyond Karel J Robot*, Joseph Bergin - Software Tools.

Bradley, Bryson, Terilla; *Topology: A Categorical Approach*, MIT Press, 2020.

Cheng; *The Joy of Abstraction,* Cambridge University, 2023.

Hare Games; https://en.wikipedia.org/wiki/Hare_games.

Lawvere; *An Elementary Theory of the Category of Sets* (Long Version) with Commentary, *Reprints in Theory and Applications of Categories*. 11: 1–35.

Milewski; *Category Theory for Programmers*, 2019.

nLab; Arrow Categories (March 22, 2023), https://ncatlab.org/nlab/show/arrow+category.

Walters; *Categories and Computer Science*, Cambridge University, 1991.

Additional Resources:

Epelbaum, Trisal; *Introductory Category Theory* Notes (PDF), https://web.math.ucsb.edu/~atrisal/category%20theory.pdf.

Holmes; *Category Theory* (PDF), 2019, https://web.auburn.edu/holmerr/8970/Textbook/CategoryTheory.pdf.

Mac Lane; *Categories for the Working Mathematician*, 2nd Edition, Springer, 1998.

Index of Terms

Made in United States
North Haven, CT
05 November 2026

21771719R10054